Mathematics for Computer Students

Rex Wilton

MANCHESTER • OXFORD

British Library Cataloguing in Publication Data

Wilton, Rex
  Mathematics for computer students.
  I. Title
  511.33

  ISBN 1–85554–167–X

First published in 1992 by:

NCC Blackwell Limited, 108 Cowley Road, Oxford, OX4 1JF, England.

Editorial Office: The National Computing Centre Limited, Oxford House, Oxford Road, Manchester M1 7ED, England

Typeset in 10/12 Times by The Midlands Book Typesetting Company; and Printed in Singapore by Chong Moh Offset Printing Pte Ltd.

ISBN 1–85554–167–X

For Diana, my wife, who has given constant support and encouragement, and for my son, Petroc, who has read and experimented with much of the manuscript in the ultimate consumer test.

Preface

This book has been many years in the making. During these years I have been asked to teach similar subject-matter on computing courses at all levels from school-leaving examinations to post-graduate and from initial industry training to in-service professional courses. It has become increasingly apparent that few books exist which deal with what is truly relevant to the needs of computer students. Certainly vast areas of traditonal mathematics courses are *not*.

It is equally apparent that few people writing mathematics textbooks for computer studies courses realise that for the computer practitioner the demands are for real, robust programs which work and for real systems which may be efficiently planned, costed, loaded and maintained. These systems must be able to deliver the very wide range of information types and quantities which the users need. The newcomer to the industry may find him- or her-self contributing to the programming of computer-controlled machine tools, aircraft flight simulators or 'traditional' commercial data-processing. Equally he or she might be working as a member of a team of operators within a large main-frame system or as the sole systems/programmer/operator 'expert' in a small business system. There are, in all these, common mathematical concepts and processes; a virtual common 'core' of essential mathematical understanding and skills.

From the moment of starting to prepare for joining the industry, the new recruit has quite specific needs which can be met from the great toolbox of mathematics. Ironically, all the elements are to be found within any modern mathematics syllabus in the schools. Much of so-called 'modern' mathematics is, indeed, in the syllabus precisely *because* of the development of computers. What is so often lacking is any meaningful connection between the teaching in schools and the *practical needs* of the real world of computing.

In this book, I have attempted to bring together the computer practitioner's needs and the mathematics concerned – to provide the connective tissue, as it were. I have ruthlessly excluded any mathematics which does not have a clear and direct relevance.

Whilst the book has been written for post-school, entry-level computer courses such as the NCC's Threshold Course and its International Diploma, the BTEC Diploma and similar initial studies, there will be much which is relevant for students on other courses. In particular, I have tried to write a book which may do something to help the many people who wish to work in computing but who have had little prior success with mathematics.

For those of us who need to use mathematical tools in practical ways the subject should not be shrouded in mysticism nor should we be 'process-bound'. Equally, I believe that ordinary users should not be subjected to 'symbol-shock' whenever they open a book on the subject. Users have been making ever-louder demands for 'user-friendly' computer systems. We, as important users, have just as much of a right to a user-friendly mathematical toolbox.

There is no such thing as a totally original book on any subject and I must acknowledge my indebtedness to pupils and students, spread over thirty years of courses, who have helped, by their experience of difficulties, concentrate my thinking about both the mathematics and the teaching methods. Numerous colleagues, in professional computing and in teaching, have also kept me company in riding my hobby-horse. George Penney and Rob Roseveare at NCC, Dr. John Roberts at NCC Blackwell and especially a former and much-respected teaching colleague, Alan Shaw, have all contributed in different degree to the development of the ideas embodied in this book. Needless to say, they are emphatically *not* responsible in any way for any of its imperfections or omissions!

This book is not in any way intended to be a mathematics course. It is rather a topic-by-topic exploration of a specific range of mathematical ideas and techniques and how and where we find them applied in our industry.

There are exercises within the text as well as at the end of each chapter but, if the topics are truly to be mastered, plenty of practice within real applications is the only way which I know of developing true confidence.

Rex Wilton,
December 1991.

Contents

10   Data structures

11   Numerical and algebraic methods

Introduction

Mathematics means many different things to different people. Many people use highly specialised maths in their chosen trades or professions but few are *professional* mathematicians. We all use some elementary mathematics in everyday life, yet quite excellent computer people may come from non-mathematical disciplines. What then are the *mathematical* needs of the computer student?

It is the author's belief, based on many years of working with both computers and mathematics, that a genuine understanding of a range of mathematical ideas and some competence in selecting appropriate mathematical processes together provide a key element of good computing practice. This applies equally to the operations staff, seeking to establish the reliability and the true capacity of their systems; to the programmer who wishes to write efficient, effective and easily maintained software, or to make best use of existing packages without having, each time, to 're-invent the wheel'; and to the analyst, seeking to devise logical, efficient and cost-effective systems.

Many people, often as a result of early failure and poor teaching, have a deep distaste for mathematics, may even fear it. If you are one of those people, do not be frightened off just yet. Almost all of us have, in our natural intelligence, a capacity for mathematical thinking and practical competence which may, sadly, just have been 'blocked off' by unhappy early experience. In this case, it takes a great deal of courage to start to look again at the subject. The hope of the author is that even the so-called non-mathematicians *will* take such a look, in the interest of becoming better computing practitioners.

Computing *was*, in earlier times, dominated by mathematicians and scientists merely because they, for centuries, had sought to devise large and fast calculating machines to help them in what most of them detested in any case, the drudgery of elaborate calculations. More recently, when electronic machines emerged as the most likely candidates for the job, they were so complex and so temperamental that they became as much the preserve of the scientist as the mathematician. The former were, of course, also in the business of having to perform vast numbers of elaborate calculations. If truth be known few people, even among mathematicians and scientists, have ever actually enjoyed 'doing sums'. It is a very tedious, low-grade activity to which the machine is much better suited than are people since *it* is very fast and rarely makes 'mistakes'. Above all, the machine never gets bored!

The real change came with the evolution of the transistor into the amazingly reliable 'chip'. The machines are now comparatively easy to use and so reliable that anyone can play! Moreover, we can 'play' with physically small machines which have vast memories, high processing speeds and powerful software. These same machines, however, remain limited by their own fundamental characteristics.

It is in understanding those limitations and exploiting the inherent characteristics that some mathematical 'know-how' is invaluable.

It was fashionable at one stage to talk of computers as 'electronic' brains but it is safer to think of them as 'high-speed morons'. They will do the wrong thing as readily as the right, introduce error where there was none and do the stupid as readily as the seemingly intelligent. They are ignorant of right or wrong, good or bad, accurate or inaccurate. *We* must supply these things and monitor the activities of the machines at all times.

Essentially, the machine consists of myriads of switches, any one of which can only be 'on' or 'off' and this establishes what the machine can do. If 'on' represents a '1' or 'true' and 'off' represents a '0' or 'false' the machine can have a set of such switches arranged so as to **add numbers written as 1s and 0s** and to **compare statements as 'true' or 'false'**. This is *all* the machine *can* do at its very core in the **Arithmetic and Logic Unit**. (Incidentally this is not the noun, 'arithmetic', meaning sums, but the adjective, *arithmètic*, meaning *by the process of addition*).

All other processes, such as multiplication and subtraction, are performed by variations of addition. Such procedures as 'sorts' are performed by both logical and numeric comparisons.

Fortunately, there already existed 'two-state' number and logical systems among the interesting but very specialised playgrounds of professional mathematicians. These met practically all the new demands created by programmable computers and little further mathematical invention was needed in order to make the two-state transistor into an effective basis for the modern digital computer.

For us ordinary mortals the difficulty is that *everything* which we want the computer to process for us has to be converted, under very strict mathematical rules and laws, into this two-state arithmetic or 'strict' logic. The converted data must then be processed, also under largely mathematical rules, to produce the information which we need. The conversion to binary arithmetic is, these days, done for us by the interpreters or compilers.

We, however, must establish the strict logic of our systems and software and ensure that the processing is done according to the rules. There is no help for it, we *must have* some grasp of mathematics!

Another problem is that the machine will introduce error in the process of conversion from everyday numbers to their binary equivalents. For example, 2.6 is an exact amount in our 'ordinary' numbers but cannot be translated into an exact amount in binary. Furthermore, most of our numbers are far less 'exact' than we might think and in the vast volumes of data processed and the very high processing speeds in the computer such inaccuracies can be generated and propagated very quickly and very widely unless we control the processing.

The final and by far the most important, difficulty is that the computer cannot and does not process reality. The real world must be converted into numerical and logical descriptions which *can* be processed or stored in binary form. Mathematics is a truly international language for *describing* many aspects of the real world. Its symbols and conventions are virtually standard throughout the modern world. At the heart of every computer program is a representation of some aspect of reality in this language. We are accustomed to such representations of reality in many other contexts; we call them 'models', small imitations of the real thing. Mathematical models *are* sometimes made of card, paper, plastic or metal like model ships or aeroplanes. Most commonly, however, they are *abstract* models constructed of symbols and there are long established processes both for constructing and for manipulating them.

This book is devoted to taking a fresh look at mathematics as it relates directly to good computing practice. We shall consider, step-by-step, the skills of creating the mathematical models, look at the existing processes for manipulating those models, numerical and logical, and at the special demands imposed by the machine's two-state systems. We shall also be concerned, throughout, with the control of the processing and the safeguarding of acceptable levels of accuracy.

1
Introduction to Modelling

Objectives

By the time you have worked through this chapter you should be able to:
- understand the basic principles of modelling;
- construct numeric, symbolic and spatial models;
- understand the operation of the laws of mathematics;
- see the place in computing of logical and statistical models;
- construct simple models in pseudo-code.

1.1 Numerical Models

1.1.1 The simplest of all mathematical models are built up with ordinary numbers which describe the NUMBER or the SIZE of something. If we look, for example at a herd of cows in a field we may see simply a pleasant country scene or, if we are farmers or scientists, a complicated picture in which there will be many, or even millions of different things, depending upon the level of detail at which we need to investigate. These things, if we mean to keep them to ourselves, we may never even need to put into words.

If, however, we need to communicate or store details of what we are seeing we must describe what we see in some way. Just how we describe it will depend upon *what* we need to convey and *to whom*. To an artist or an author this could be a very complex task but he or she will find it impossible to represent everything in a scene and will select what is essential to the story or the picture and to creating any necessary 'atmosphere'.

1.1.2 The process of mathematical modelling is similar but even more selective. It is a process of continuing to eliminate *all irrelevant* information, no matter how interesting it may be, until we can simplify our description no further and no possible ambiguity remains. For a mathematical model we might simplify the present example by making it absolutely clear that we are concerned *only* with the cows and with nothing about the cows except 'how many?'. The simplest possible model is then '10 cows'.

1.1.3 This number itself, as written in figures, is a model developed at a much earlier stage of man's history when he invented the numerals 1,2,3,4,5,6,7,8,9 to record the number of fingers upon which he 'counted' the cows, and the cypher, 0, to indicate when and where there is nothing (or no cows). Supplement these numerals, or figures, with a column system and we have a model-making set which can describe, in the simplest possible way, any quantity, no matter how large or how small. All that we need in addition to the numbers is, in everyday life at least, a label to show 'of what?'. The number 10, for instance, is a purely theoretical quantity until we attach to it, in our example, the label 'cows'. The label is vitally important, not only in denoting what we are 'enumerating' (giving a number to or counting), but because, as we shall see later when we consider error, the 10 when applied to cows is significantly different from the same number in *10 metres*.

1.2   Symbolic Models

1.2.1   If we advance one step further and contribute some symbols to denote the operations which we can *perform* upon numbers, +, −, ×, ÷ , (commonly called 'operators') and a symbol for equivalence, '=', we are able to make much more useful models which are only slightly more complicated. (In computing practice we have come to use '*' instead of '×' and '/' in place of '÷' and this convention will be followed in this book.) We must also be aware that the '+' and '−' symbols each have two meanings. They may be used as an *instruction*, or operator, as in 4 + 2, or 4 − 2. Their second use is in *describing* a number value as in +2 or −2, that is to say 2 greater than zero and 2 less than zero respectively. In terms of language the first use is as a verb, or 'doing' word; the second is as an adjective or 'describing' word. Some authors have tried using the signs in a different position to indicate the adjective, +2 or −2 but whilst this does emphasise the difference it has not become general practice.

1.2.2   The manipulations which we may perform with numbers and operators are not, as many people think, new tricks which have to be learned for each different application. They are always consistent and are governed by a very small number of laws. Since these laws must be observed in all our modelling we should study and learn them thoroughly. We will add a few others to the list as we meet the need for them but the majority of our processes are governed by the first ten, which are listed below. (In all cases we are using a, b and c to stand for any quantities we may choose to replace them by.)

1.3   The Laws

1.3.1   1a) $a+b = b+a$ 1b) $a*b = b*a$
          2a) $a+(b+c) = (a+b)+c$ 2b) $a*(b*c) = (a*b)*c$
          3    $a*(b+c) = (a*b)+(a*c)$
          4a) $a+(-a) = 0$ 4b) $a*(1/a) = 1$
          5a) $a+0 = a$ 5b) $a*1 = a$
          6    $a * 0 = 0$

1.3.2 You should practise replacing the letters with number values in each of these and satisfy yourself that each law always works. You may have noticed that the operations of subtraction and division do not figure in the list; why?

It may well be the first time that you have seen the laws set out in this way, although some may look very obvious and you will have been obeying them for years if you have been getting the right answers to your calculations. If you have been prone to getting a lot of answers wrong, it may be because you had no idea that such laws existed and have been doing things a bit haphazardly. In any case we should have a little more knowledge than just the bare list, so let us give the names of the laws and their interpretation.

1.3.3 1a) and 1b) The COMMUTATIVE laws for addition and multiplication.

The word 'commutative' means 'in both directions', as in the word 'commuter', one who travels *to* work, in one direction, and *back from* work in the opposite direction. Mathematically it means that if we are *only adding or multiplying* a set of numbers (no matter how many, even though only two are shown above), the sequence in which we add or multiply those numbers does not affect the result. Hence one traditional way of checking our working if we have 'added up' a column of numbers is to add them 'down' and an equally effective way of checking 7 * 3 is to do also 3 * 7. In practice, it also means that we may put such a sequence of additions or multiplications into a model in the order in which they are most easily understood even if this means re-arranging a traditional way of writing the model.

Check for yourself what happens with subtraction and division. Are they also 'commutative'?

1.3.4 2a) and 2b) The ASSOCIATIVE laws for addition and multiplication.
This is a variation on the sequencing of operations which, among other things, allows us to split a sequence in different ways if it is easier to do so or if it helps confirm the accuracy of what has been calculated. For example 15 + (17 + 5) may be easier for us as (15 + 5) + 17. Once again you should practise picking out the easier associations of numbers. Try testing the law to see if it is equally true for subtraction and division.

1.3.5 3 The DISTRIBUTIVE law for multiplication.
This is a little more complicated but it tells us that if we are to multiply a number which is made up of two or more other numbers added together we can multiply each of the parts and add them all together at the end. It is this law which governs our 'long' multiplication. 7*432 could be spoken as 'seven times four hundred and thirty and two' or 7*(400+30+2) and we do, in fact use the commutative law to change it to 7*(2+30+400) and perform it as (7*2)+(7*30)+(7*400). We could always still set out such calculations in this way but we have, over the centuries, also introduced some 'carry' procedures which make the whole thing look rather more complicated.

A combination of laws allows us to perform 297 * 463 as

        (200 + 90 + 7) * (400 + 60 + 3) or
               7 *   (3 + 60 + 400) +
              90 *   (3 + 60 + 400) +
             200 *   (3 + 60 + 400).

Explore this and similar examples and try to identify which laws govern which stages of the whole operation.

1.3.6 4a) and 4b) The laws of INVERSES for addition and multiplication.

These, although they may seem a little strange at first, actually do crop up in many applications and may be explained in two slightly different ways. For addition:

i) for every number ('plus' number) there is an exact opposite (negative or 'minus') number which gives the answer 0 if the two are added together, *or*, if the sum of two numbers is 0 the numbers must be inverses of each other.

ii) If we add to any number its own (additive) inverse the result is zero. $2 + -2 = 0$. For multiplication:

i) for every number there is another number which generates the answer 1 if the two numbers are multiplied, *or*, if, when two numbers are multiplied the answer is 1 the numbers must be inverses of one another.

ii) If we multiply inverses the result (product) is 1.

Following on from these two laws we have a very important development. We can achieve the same result, for any addition or multiplication, by performing the 'opposite', (inverse) operation with the inverse value. For instance: $2 \div \frac{1}{3} = 2 * 3$. Similarly $4 - (-2) = 4 + 2$. This is a process which we use very frequently in simplifying models.

1.3.7 5a) and 5b) The laws of IDENTITY for addition and multiplication.

Once again these two laws may look so obvious as to need no discussion, but we do use them extensively in modelling and the idea of something 'retaining its identity', ie remaining unchanged, is one upon which we often rely. Briefly, we can sum up these laws as meaning that if the effect of any operation we may perform, no matter how complicated, is to add nothing to a number, or to multiply that number by 1, the number will remain unchanged.

In simplifying fractions by 'cancelling' for example we might, from 28/44, derive the equivalent form of $\frac{7 \times 4}{11 \times 4}$ which we can then re-write as $\frac{7}{11} \times \frac{4}{4}$. Since the $\frac{4}{4}$ is, according to the law of INVERSES, worth 1 and since, by the law of IDENTITY, multiplying by 1 leaves a value unchanged we are able to say that the true value of $\frac{28}{44}$ is $\frac{7}{11}$.

1.3.8 The law of MULTIPLICATION BY ZERO.

Of all the laws this will seem to many people the most obvious. No matter how many lots of nothing one has, it is all still nothing. Yet, in our modelling, we do often contrive to write models which include a multiplication by zero. Attempting, within a program, to *divide* by 0 is a common enough mistake to have justified including in many compilers and interpreters a specific error message, 'division by zero'. There is rarely such an error message for a *multiplication* by 0 yet, when we are debugging programs, such mistakes, sometimes catastrophic, are not uncommon. This very obvious law is unintentionally broken all too often!

1.3.9 As a way of exploring the influence of the symbols and the laws in modelling let us return to our original example. If someone opens the gate to the field and allows three of the cows to wander off we might describe the situation by saying, "From our original group of ten we have allowed to escape, (removed), three and this leaves a herd of seven". Or, "Ten less three leaves seven". This is fully represented by the model $10 - 3 = 7$.

There are many possible variations on such arrangements of words. If, for instance, we did not see the cattle straying but merely knew that only seven remained in the field we might say, "There were ten and now there are only seven; how many have strayed?". In this case we have an unknown quantity which we need to find and our model might well be $10 - ? = 7$.

It is, in practice, very common for our models to contain unknown quantities which we need to find and we usually do two things to simplify the discovery process. First, we put a stated symbol in place of any question marks and then we arrange our model so that *all that we know* is grouped together and *what we need to find out* is clearly identified. If, in the present case, we stipulate, LET (us agree that for this purpose that) 'S' (will) represent the (unknown number of cows which have) strayed, then our model will be $10 - S = 7$. Here, however, the 'knowns' and the 'unknown' are muddled together but if we rephrase the description as, for instance, "The number of strays is the difference between the original number of cows and the number still in the field" we can rewrite our model as $S = 10 - 7$ and hence, if we remember our 'subtraction facts' and we know that the difference between ten and seven is three, we can solve our problem.

We should not think it worth the effort of writing a computer program to solve a single problem of this type since it would take more time than we should ever save. It is really only worth such an effort if we use the computer to deal with many repetitions of solving precisely similar problems using the same model, or with very complex problems. If we were called upon to deal very frequently with simple 'difference' problems we should devise a 'generalised' model where the numbers concerned could be input from the keyboard (or other input device), or from a stored file.

1.3.10 It is, nevertheless, very valuable to spend time thinking about and practising the building of such simple models because every mathematical statement has been derived from, and can be re-translated back into, descriptions in our ordinary language. Much of the technique and many of the methods for doing this is what we have learned earlier as algebra. In fact the very word 'algebra' implies 'writing down complete statements'. In practice, we simplify such statements by replacing words with symbols wherever possible. Many of us, however, do not feel too comfortable dealing with symbols only as we are, quite naturally, more at home with words. Mathematicians, on the other hand, know that words are actually very slippery things and may be interpreted very differently by different people. In order to avoid any possible ambiguity we use either internationally agreed symbols or we state what our symbols are to mean in any particular case. Symbolic statements can also be very brief and this is often a further help to clarity. Throughout this book we will discuss the meanings of the symbols we use, as we meet them, and try to make them as 'user-friendly' as possible.

1.4 Generalised Models

1.4.1 There are five words which become important immediately we start talking about the next stage in modelling for computing: ASSIGNING, CONSTANT, DECLARING, PARAMETER and VARIABLE. One problem for many of us when reading programming manuals is that we are not always comfortable with the language used and these particular words are common stumbling blocks. This is partly because at least some of them are rather different from what we meet in traditional mathematics. We will try always to discuss such differences of usage. (It is also perhaps confusing that in computing we make use of these same words when dealing with alphabetic, or 'string' data. This confusion is made worse by the fact that all characters, alphabetic as well as numeric, are actually stored and manipulated as number codes. We, however, are only concerned with the true numerics and the mathematical symbols in this book.)

1.4.2 Let us examine the idea of a CONSTANT first. If, for instance, we consider the comparison between the diameter and the circumference of a circle we find that all the evidence, gathered over many centuries, indicates that the relationship is fixed. The *larger* the diameter the *larger* the circumference; not in any haphazard way but always with the circumference 'three and a bit' times as long as the diameter. Such a fixed relationship we describe as a 'constant', in this case labelled as 'pi' or 'π' simply because they were the labels used in the original Greek explorations of that relationship and have been accepted as standard 'labels'. There is a special difficulty with π since we have as yet no fixed value for it, although mathematicians still have great fun computing values for π to hundreds of thousands of places of decimals. We may, therefore, use a value for this constant which is appropriate to the task being performed. That value will, however, be written into any program which uses it and will remain the same throughout that application, just as we should use a consistent value for it in any set of manual calculations.

1.4.3 VARIABLES need both a name and a value and whilst the name, once decided, will be fixed, the value may well change, more or less frequently, during operations on the model. Variables will *not* usually have universally agreed labels and each must be given one. Within the programming team, we must agree upon both the label and its interpretation. We must also be sure that each variable name is unique so that no ambiguity or confusion may arise. The chosen name will often be a single letter in traditional mathematics but may be a complete word or phrase in modern programming languages. The label must be used consistently throughout the model. In our example the word 'COWS' is the variable name. In programming we must DECLARE all such labels at some stage, that is to say that we must include the labels in the program and the documentation. (In good programming practice we shall also record them in the data dictionary.)

It is then common practice to ASSIGN an actual value, or PARAMETER, (meaning only a quantity or value) to that variable either within the program or by input from keyboard or file. This means that we shall store, in memory, at the address allocated to the variable which we have declared, the value which we intend it to have at any given stage of the program run. (Once again, in our example, the variable 'COWS' is assigned the parameter 10). If the program is to be used for a number of runs we must also, as a matter of good practice, initialise all variables to zero so that the processing is not corrupted by any random value stored at the relevant address or by a value remaining from an earlier run.

1.4.4 To illustrate all this let us make a generalised model of the 'missing cow' problem. First we label and declare the variables: LET (and this is a classical mathematical word for declaring variables), F=the numbers of cows in the field to start with; (LET) R=the number of cows remaining and (LET) S=the number of strays.

The next step is either to ASSIGN or INPUT parameters for each of the 'knowns'. Thus either:

LET F=10; F←10 (if we are assigning within the program),
        OR   INPUT F (if we are taking keyboard input).
LET R=7; R←7   *OR*   INPUT R.

(The left arrow, ←, is a common symbol for assignment; whatever is on the left of the arrow takes on the value of whatever is to its right.)
These steps together complete the process of establishing the 'knowns' and

LET S = F-R; (S ← F-R); completes the processing.

We now need only 'PRINT S' to convey the answer to us.

1.4.5 This is the essential core activity of all programs; *Operating as needed upon a model of the problem which the program is designed to deal with*.

1.4.6 Our memories of school mathematics will call to mind a variety of models which we were required to learn. $A=L*B$ is wholly made up of variables. $V=4/3(\pi r^3)$, on the other hand, has only one variable from which the unknown, 'V', is eventually derived; can you identify all the constants? As a useful task write down some of the models which you can remember and try to identify which are constants, which are unknowns and which are variables. Then try rewriting the models in as many different ways as you can.

If we take $A=L*B$, for example, we can rewrite it as $L=A/B$ or $B=A/L$ so that, eventually, whichever two parameters are known we can derive the third, unknown. Try creating such variations of the $V=4/3(\pi r^3)$ model.

1.4.7 One task which frequently falls to the programmer is to take an existing model and adapt it to accept some new arrangement of variables in order to solve the problem for a different unknown. Do not rely on remembering such models. It makes much more sense for the programmer to know where to find such them in the reference books. It is also a waste of much energy to re-invent a model which someone has developed perhaps centuries earlier.

1.5   Spatial Models

1.5.1   Many applications of computer processing are concerned with design, manufacture or movement in the real, three-dimensional world. To model real objects,
real places and real journeys we need something more than simple numbers or
symbols and we turn to graphical representations and various geometries for our
tools. The VDU screen and some other computer output devices are powerful means
of *presenting* spatial information but our *processing* must depend upon numerical or
logical representations of the three-dimensional world of space and time, since the
ALU is limited to these two types of operation. We have, traditionally, explored and
represented such realities by technical drawings, graphs and maps. For the computer,
however, whether the problem is one of automatically piloting an aircraft across
oceans or causing the cutting edge of an automated machine tool to drill a cylinder
block for a motor car, we have to make computer models which adapt or replace
these traditional mathematical processes.

1.5.2   The primary tools for building such models are the co-ordinate descriptions,
bequeathed to us by the great French mathematician, Descartes. The data from such
descriptions we may organise in the matrix or array and its simpler relation, the vector.
Once again we shall find that, if we can construct the models, we shall be able to select
from existing mathematical processes in order to perform the necessary operations
upon them.

1.5.3   In this book we shall practise with two-dimensional problems only but
everything which we consider may be extended to three or more dimensions by
comparatively simple extensions to the basic models and the techniques for their
manipulations. It will often be helpful to see a sheet of graph paper itself as a model
of the VDU screen and vice versa.

Descartes initiated the great conceptual leap of setting realities, both spatial and
numerical, on to a square grid of which each axis was a number line, itself a model
of our number system. You can see this in Figure 1.1.

Figure 1.1   The number system as number line

1.5.4   When two axes, at right angles to one another, are each designated by such
a number scale and intersect at 0, problems involving two numeric variables or
two-dimensional space may be modelled. Any position in the space denoted by the
grid may be uniquely identified by stating two numbers, the first representing the
distance of the point from zero along the horizontal axis and the second its similar
distance along (up or down) the vertical axis. This can be seen in Figure 1.2.

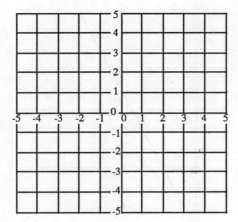

Figure 1.2 The grid devised by Descartes

1.5.5 Provided that the order of the two numbers is always the same each such 'ordered pair' describes a precisely identified, unique, point. (See Figure 1.3.)

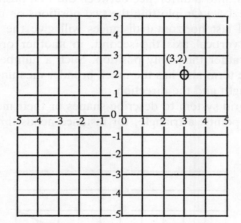

Figure 1.3 The point as ordered pair

Two or more such points may define a line. You may be familiar with the description of the relationships between two variables by the line on a graph plotted from such points. This is shown in Figure 1.4.

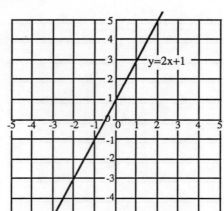

Figure 1.4   Y = 2X + 1 as a graph

This also gives us exactly what we need for our spatial modelling, a simple means of converting a position, or positions, in space into a purely numerical description.

1.5.6   When we move from one such point to another we can simplify that journey, in purely spatial terms, into a difference between our first position and that at which we finish. (See Figure 1.5.) On plotting the two positions on our grid the *difference* between the first and last horizontal distances will give the first number and the difference along the vertical axis the second, in another ordered pair; this time denoting movement rather than only position. Such a number pair is described as a vector, although the term also has the much broader meaning of any line of which we know both the length and the direction.

When we use our grid system to describe shapes or their movement we may well identify a number of points; a triangle, for instance needs three, as can be seen in Figure 1.6.

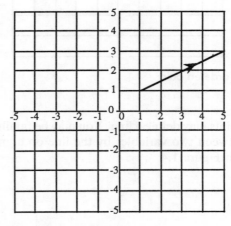

Figure 1.5   The journey from (1,1) to (5,3) = (4,2)

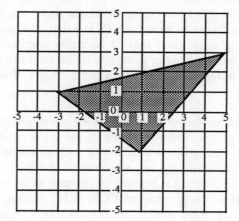

Figure 1.6 A triangle as ordered pairs

1.5.7 Each of such points will have its own ordered pair. These ordered pairs will themselves need to be put into and maintained in a fixed sequence and an existing mathematical model, which has ready-made rules and operators for processing, is the matrix or array. As a matrix, purely numerical model, our triangle may be stated as:

$$\begin{bmatrix} -3 & 1 & 5 \\ 1 & -2 & 3 \end{bmatrix}$$

The mathematics of matrices existed before modern computers but they might well have been made for each other, so well adapted is the machine to the manipulation and processing of matrices. Under the name of arrays they have evolved into one of our most flexible and powerful data structures.

1.6 Logical Models

1.6.1 One part of mathematics which is often taught very badly is the application of logic. Even students who have been taught 'sets' rarely understand the relevance or the rigour of what they have learned yet, of all the fundamentals of mathematics, this study ranks as being of the very greatest importance. We shall examine this topic in greater detail later but let us try to establish the basic principles.

It is an age-old principle that we can only logically compare *like with like* so that it is, for instance, quite ludicrous to make statements such as "She is more beautiful than he is tall." It would be equally ludicrous, but even more of a trap, to say "On a scale of one to ten for beauty she rates 8 whilst he, on a similar scale for tallness, rates only 6." Can you see and explain why? One of the great problems of mathematical thinking is to decide just what may properly be described by a measurement and what is, and always will be, just a matter of opinion.

That division is between PARAMETRIC (attributes which can be measured or given *parameters*) and NON-PARAMETRIC mathematics and modelling.

1.6.2 It is also of fundamental importance to realise that any measurement of one characteristic very rarely tells us anything whatever about any other characteristics of the items which we are trying to compare. To say that one person weighs more than another tells us nothing more than that; it certainly does not justify, for example, inferring or implying that the heavier person is taller, or more cheerful. For this kind of modelling we have to learn to think much more clearly, and in a more disciplined way, than that.

1.6.3 To start with we shall need to consider logical relationships between objects, numbers, statements, people and everything in the universe which may be logically compared. Instead of writing out a long list of such possibilities each time, we use the word 'ENTITIES' to stand for any single item of whatever kind it may be. An entity, then, is anything which has a logical existence of its own. When an entity is included within a set it is described as an ELEMENT of that set.

Entities, in turn, may have many different features, qualities, measurements, characteristics and so on – yet another long list of the multitude of different aspects of any entity which may concern us. The word we use to avoid writing the list every time is 'ATTRIBUTE'. The final word, nowadays in quite common use in this context, is 'SET'. It is often thought that any collection of entities is a set but this is not so. It is of the greatest importance, before embarking on any study of logical relationships, to be quite clear as to what constitutes a *mathematical* set.

1.6.4 The rules are brief but very clear:

1. A set must be DEFINED. In this case the word has a much more rigid meaning than usual. The set, whatever our purpose in using it, must be so clearly defined that we can say, of any entity, that it belongs either IN or OUT of the set; there *cannot be any 'maybe's'*.

To take an example, we cannot have a *mathematical set* of 'people with blue eyes' since there would be endless arguments as to whether one person's eyes were blue or grey or another's were blue or green. If it is possible to argue whether an individual entity should belong in a 'set' or not it is not a true set. This is the real meaning of 'two-state logic', which is all the computer can truly deal with in ordinary practice. Is there a set of tall people? You should, with a partner, 'define' some sets and then test them to see whether they truly conform to this rule of definition. Do this carefully and test your definitions against some extreme cases. It is thus that we create test data which will truly expose some of the logical errors which cause our programs not to run successfully or reliably.

2. Every element in a set must be unique. There are no duplicates in true sets. This immediately rules out many of the sets we recognise in everyday life. It may be entirely sensible, in ordinary life, to talk of 'a tea-set', for example. It cannot, however, be a *logical* set since it contains, say, six identical cups, six identical saucers and six identical tea-plates. In the equivalent logical set this would be reduced to just one of each of the items. Furthermore if we put together, for example, the set of numbers, 2, 3, 4 & 5 with the set of numbers 4, 5, 6 & 7 we would have the set 2, 3, 4, 5, 6 & 7. This is one of the situations where 4 + 4 does not equal 8 since we are adding two sets of four items together but have a set of only six items at the finish. Once again you should try a number of such 'additions' to be sure that you understand what is happening.

We are actually quite familiar with these ideas. In retrieving and sorting records in our computer files the allocation of unique 'key' fields plays a vital part. The combining of sets is what happens when we merge two files. What would happen during a merge if two records in the originally separate files had the same number in the key field?

Strictly speaking any filing system, including traditional manual ones, must abide by these rules if we are not to lose track of items in those files. Many program failures stem from inadequate understanding of the importance of uniqueness.

1.6.5   One common misconception is that all the elements of a set must have an obvious attribute in common. This is not a true test since we may group entities into any set which suits our purpose, so long as it obeys the rules. If I wished to do so I could define the 'set of things in my left trouser-pocket'. It would, as I write this, consist of a bunch of keys, a handkerchief, a pocket-knife and a small wallet. The only attribute those things have in common is that they are, at the moment, in the relevant pocket! Such sets may not be especially useful in routine D.P. but they are entirely valid nonetheless.

1.6.6   As other logics are developed the situation is changing slowly, in applications such as 'artificial intelligence' for example, but for all conventional information processing our databases, files, records and fields must conform to these two-state conditions.

1.6.7   Few interpreters or compilers can deal directly with the notation or the manipulation of sets. It is we, the human beings, who have to learn to think in this way, to allocate material to such logical structures and to plan and program the manipulation of information in such structures. The one exception is with a database management system, DBMS, which operates strictly on logical sets. Even here, however, it is we who have to make sure that the two-state logical interpretation of the information is properly established.

As with all mathematical modelling there are consistent symbols and notations to be mastered if we are to be truly fluent in interpreting such structures.

1.6.8   One of the most practical is the Venn diagram (shown in Figure 1.7), by which we can show quite complex relationships but which develops from the relatively simple notion that we can define a set by 'fencing it in' within a circle (or an ellipse), so that any entity outside the circle is a 'not' and anything within is an 'is'. We can then combine circles to express very complicated relationships within and between sets. Any entity may belong to many different sets. Try making a list of the sets to which *you* belong.

Test them very carefully for definition but you may find that the 'uniqueness' criterion is less of a problem. Why is this?

In order to limit the complication of what we are trying to organise into sets, and any 'family tree' of sets is known as a TAXONOMY, we first define the whole area of our problem. This was known, in more formal times, as the 'universe of discourse', the actual and limited world we are examining or discussing. This has come to be described, a little more simply, as the 'universal set' and in Venn diagrams is usually denoted by a rectangle surrounding all the sets in which we are interested. A company's stock list, for example, may be seen as one 'universal set' which would contain many other sets.

Finally, a set which had no relationship with any other set would be a very strange one indeed. It is the complexity of how sets relate to one another and how these relationships may be expressed in the simplest possible way as 1s and 0s which we shall need to study in detail.

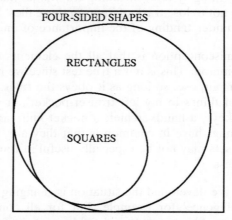

Figure 1.7   A universal set and two of its 'inner' sets. Can you interpret the story which they tell?

1.7   Statistical Models

1.7.1   There are many situations where 'ordinary' numerical mathematics cannot provide all the models or tools which we need.

In very complicated situations, such as describing a whole country, actually counting or measuring *everything* about that situation would take too long, cost too much, need too many people or pose other very complicated measuring or counting problems. We sometimes describe such situations as COMPLEX STATES and we can usually deal with them only by taking samples and using those to give us insights into the whole complex. Because of the vast numbers of operations, data items handled and stored, and the sheer speed at which it operates, a computer system is actually just such a complex state and we are dependent upon statistical models and methods to plan and monitor our systems, their loading, reliability and performance.

1.7.2   The second area of interest which can only be modelled statistically is the situation where complete testing destroys what is to be tested. Manufacturers of light-bulbs, for example, are often required to certify and guarantee the life expectancy of their products. No-one would buy such things from a manufacturer who could not assure us of a reasonable period of reliability yet the only way to know for certain how long a light-bulb will last is to keep switching it on and off and 'running it' until it fails.

It would, however, be quite absurd to offer to sell such a lamp with the guarantee that 'We know that this lamp will burn for 1500 hours because it has already done so'! We expect reasonable guarantees of life expectancy and reliability from most things we buy, from electric razors to motor cars and aeroplanes; especially aeroplanes! Yet here actual measurements of tests to failure are useless to us and only statistics allow us to model the problems.

1.7.3 Yet another area is where, based on the recorded history of past events, we need to be able to make reasonable forecasts of the future. Weather forecasting is just one very obvious example and again the modelling tools are statistical.

In weather forecasting the models are astonishingly complicated and the corresponding computer programs extremely complex. In these and similar areas we are much concerned with 'trends' or 'tendencies' and trend-analysis is a powerful management tool.

1.7.4 Finally, although we shall not delve into non-parametric statistics in this book, there is the analysis of opinion. The marketing of products is just one field where statistical models form a major source of management information based on the evaluation of opinion.

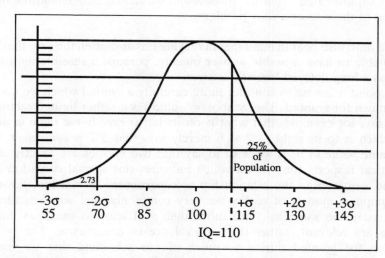

Figure 1.8 One of the commonest statistical models. Many human characteristics are distributed in this way

1.7.5 What all these aspects of the real world have in common is that in all of them, by their very nature, we are denied the certainty of precise measurement. What our statistical models and processes allow us to do is *to diminish the extent of our uncertainty*. One of the saddest spectacles of modern applications of mathematics is that so often the 'persuaders' ignore that this is the fundamental nature of statistics. They try instead to delude us by saying " . . . statistics *prove* that . . . ". This is a manifest untruth. Statistics deals in tendencies, likelihoods and similar *un*certainties and whilst it may offer us evidence that something is more or less likely to happen, or to be the case, it can never *prove* anything much less tell us *why*.

1.8   Algorithms and Trial-and-error

1.8.1   Traditionally mathematicians have always sought to create 'algorithms' for processing their numbers and the word meant 'a fixed procedure by which, if we provide the right inputs, follow the procedure correctly and obey all the mathematical laws whilst doing the correct mental arithmetic, we will find right output'.

The 'drill' by which you were taught to perform 'long' multiplication is just such an algorithm. Algorithms relieved us of the need to re-invent processes every time we needed them. On the other hand we had to memorise them and make sure that we didn't forget any steps or how to perform them; a tall order for many of us. A properly designed and written computer program, which has been correctly tested, should be an algorithm. Right inputs = right outputs.

1.8.2   However, we have, as an additional tool, the speed of the computer and this can be put to good use in running 'try-it-and-see' models. The formal name for these is HEURISTIC models and they allow us often to avoid constructing an algorithmic model in favour of one in which we try out various combinations of values in order to see if we can find a suitable solution to a problem. Spreadsheet systems often allow very sophisticated 'what if?' models and we should understand the nature and application of these to everyday problems.

1.8.3   We must also bear in mind that in real life it is also often the case that it is much more valuable to have a usable answer quickly, perhaps a sensible approximation, rather than a long-delayed 'precise' answer.

At this point it is wise to consider more carefully a symbol which we undoubtedly take too much for granted. The symbol '=' indicates a rather limited relationship. It does not say, for example, that what is on its left *is exactly the same in all respects* as that which is to its right. $3*2 = 6$ merely says that $3*2$ is *equivalent*, that is, it has the same *value* as 6. If we wish to say that two entities are exactly alike in all mathematical respects the symbol which indicates this special condition is written as '$\equiv$' and we describe the relationship as **congruence**. In real life, however, and in our computer models of reality we very commonly deal with situations where 'approximately the same as', 'not more than', 'at least as much as' and similar conditions are relevant, rather than equivalence or congruence. The '=' sign has been taken for granted during so much of our schooling that we may find it difficult to see intelligent approximations as being just as important, as indeed are the *non-equivalence relations* or *inequalities*.

There are several symbols in use for 'approximations' but in this book we shall use '$\approx$' throughout for the sake of consistency. Similarly you may come across different ways for writing the 'not more than', '$\leqslant$', and the 'at least', '$\geqslant$', but their use in the book will be consistent. The symbols for 'more than' and 'less than', $>$ and $<$ respectively, are used more consistently and should present no problems.

1.8.4   To see our inequalities a little more clearly let us look at the graph of $y=2x+1$, shown in Figure 1.9. The line itself consists of all the *points* on the graph where the statement is true. That line is, then, the line of the *equivalence*. What of all the rest of the surface indicated by the grid? Take any point below and to the right of the line, say (2,2). Here the $x=2$ but so does the y and since the y only *equals* the x it must be *less than* $2x+1$; or, $y < 2x+1$. If we try a few more such points, even those very close to the line of $y = 2x+1$, we shall find that this whole region of the graph is described by $y < 2x+1$.

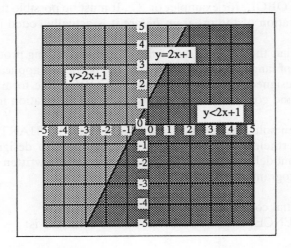

Figure 1.9 The graph of $y = 2x + 1$

1.8.5 What about taking that region and *including the line*? Can you write down the model? Go on then to look at the region above and to *the left of the line*. What will be the INEQUALITY MODEL which describes that region? What will be the model if we include the line?

Can you see that every time we state an equivalence relationship we are also implying a whole set of inequalities? Try some others of your own devising and try to get used to seeing a graph as a whole set of relationships as well as the one which we choose to 'pick out'.

Here again is a small but extremely important section of mathematical ideas where the testing of our programs is concerned. We must decide, whenever we program any equivalences, just how our program should deal with the corresponding *inequalities* and what values our test data should include in order to reveal any defects or omissions in design which have arisen from ignoring them!

1.9 Modelling in Pseudo-code

1.9.1 Whatever *mathematical model* we select, it is still only MAN-READABLE. The stage must ultimately be reached when we must make it MACHINE-READABLE by designing a program segment to translate it into computer-related form.

1.9.2 Unfortunately, each high-level computer programming language has its own ways of doing things and many even have numerous 'dialects'. There are, for instance, many variations of BASIC, but just as a good program should, ideally, be 'ego-free', that is to say in a standard style which is independent of the whims of the individual programmer, so should a good program *design* be 'language-independent'.

1.9.3   This means that the design should be practicable whether it will eventually be programmed in COBOL, for example, or 'C'. It must be possible to test the logic and the planned operation of the program *before* it is actually coded in the language of choice.

It also means that the designer, after the design and testing processes, can choose which of a range of languages is best suited to the particular application. If, however, the program is designed from the beginning in COBOL-like form and terms it may be very difficult and costly to *redesign it* for eventual application in BASIC, say.

1.9.4   The chosen solution in times past was the FLOWCHART and this still has value. It is, nevertheless, a great advantage to be able to design a more directly program-related model in PSEUDO-CODE which can be written in simple English using standard programming CONSTRUCTS.

1.9.5   The essential constructs are:
1) SEQUENCE, and in PSEUDO-CODE form this is:

> Process A
> Process B
> Process C

> END

Each part of a complete program run being completed in turn.

1.9.6 2) SELECTION, of which there are several forms. Among them we may identify several varieties of a SKIP, or BRANCHING, to another process or routine.

> 2 a) SELECTION by SINGLE IF;   pseudo-code:

> IF condition
> > DO process
> ENDIF   or:

> WHILE condition
> > DO process
> WEND

> 2 b) ALTERNATIVE IF;   pseudo-code:

> IF condition
> > THEN DO process A
> > ELSE DO PROCESS B
> ENDIF

2 c) MULTIPLE SELECTIONS;   pseudo-code:

```
IF condition 1
     THEN DO process A
          ELSEIF condition 2
               THEN DO condition B
          ELSEIF condition 3
               THEN DO condition C
                    .
                    .
                    .
ENDIF
```

Which may continue for any number of conditions subject only to any limitations in the language in which the program is eventually to be written.

1.9.7 3) There is also the LOOP, which may be UNCONDITIONAL or CONDITIONAL, SINGLE or NESTED, and which may be RECURSIVE.

3 a) A SINGLE LOOP may be, in pseudo-code:

```
read N
     DO K = 1 to N STEP 2
          process
     ENDDO
          K ← K + STEP
END
     in simple, 'loop-counter' form.
```

3 b) DOWHILE condition
```
          process
     ENDDO
```

controlled by a 'logical' evaluation of 'condition'.

3 c)   DO UNTIL condition
```
          process
     ENDDO
```

again a logical decision to end the loop.

3 d)   The NESTED LOOP   pseudo-code:

```
read N, N2
     DO i = 1 to N STEP 1
          DO j = 1 to N2 STEP 1
               process A
          ENDDO
          process B
     ENDDO
END
```

may be nested to 6 levels 'deep' in many languages.

There are some conventions and words associated with this form of modelling among which is the INDENTATION of constructs within constructs as an aid to easier reading.

1.9.8 The use of words is fairly simple because of the comparatively plain English coding. We tend to use:

READ to imply that data is retrieved from a file;
ACCEPT to indicate a pause for data input from keyboard;
DISPLAY to show output on a VDU screen;
WRITE to output data to a file and
PRINT to send output to a printer.

In addition to these, for working examples or setting up test data, we may use DATA to indicate that what follows is a data list to be READ into the program. In other words we may emulate the contents of a file.

1.9.9 If we put some of these ideas into a specimen program segment we may try our hands at identifying what happens at each line and do a pencil-and-paper test run with some different data.

```
read NAME, AGE
     DOUNTIL end of file
          IF AGE <30
                    display NAME
                    accept DEPARTMENT
                    print NAME, AGE, DEPARTMENT
                    write NAME, AGE, DEPARTMENT
          ENDIF
          read NAME, AGE from next record
     ENDDO
END
```

1.10 Conclusion

The choice of appropriate models will usually be a matter of selecting from what is already available in the mathematical sources. Whether such models will fully serve our purposes or will need modifying, or whether new models will need constructing from scratch, is a matter for careful analysis and judgement. It is we, in consultation with the users, who need to have the understanding and the knowledge to make such judgements and to select or develop models accordingly. Remember that many of those which we need will have already been developed and tested and these, if we know what to look for and where, will save us a great deal of time and hard work.

Exercises

1) Which of the laws in 1.3.1 do *not* apply to subtraction and division? Show why.

2) What variable types and parameter ranges would be appropriate to:

   a) a person's age;
   b) a person's sex;
   c) a date;
   d) a time;
   e) the price of a motor car;
   f) an hourly wage rate;
   g) the number of hours worked by an employee in a week?

3) Write a list of: a) six proper sets
   b) six groups of entities which are not proper sets.
   Give reasons in both cases.

4) List at least ten sets of which you are a member in the form:
   $I \in \{.................\}$

5) Identify five *each* of real-life situations where values must be "at least . . . ." or "not more than . . . ."

6) Write a pseudo-code model of a procedure to accept and validate a user-code and password to permit or deny access to a computer system.

2
Constructing Models

Objectives

After completing this chapter you should be able to:
- develop a model following an organised process;
- convert a specified situation into a model;
- construct models of simple commercial and other real situations;
- understand the differences between algorithmic and heuristic models;
- make models using matrices, tables, trees and diagrams.

2.1 The Modelling Process

2.1.1 The process of actually developing or constructing a model can be organised into a number of logical stages. If we then follow those steps in a deliberate sequence, we are likely to find the process becoming a good deal easier. Like everything else, it needs practice to develop confidence.

2.1.2 The development of the model may be reduced to a sequence of questions:

Stage 1 Is the problem truly one which demands a mathematical solution?
If the problem is numerical or strictly logical the answer will undoubtedly be "Yes". If the customary response to such problems is simply an *opinion* the answer, in all probability, will be "No".

Stage 2 What *relevant* facts are known?
(When we first encounter a problem the biggest difficulty, often, is that there is just *too much* 'information', which may simply confuse the issue.)

Stage 3 What other facts do we need and where can we find them?

Stage 4 Does the problem fall into stages and is there a *necessary*, or a *desirable*, sequence?

Stage 5 Can the facts be represented by symbols and separated into 'knowns' and 'unknowns'?

Stage 6   Is there an *existing* model which will serve, or may be adapted, in order to save us the labour of inventing a new one?

Stage 7   Is the existing model one which makes good use of the capabilities of the computer?

Stage 8   How should I express my model in terms which the programmer can understand?

2.1.3   When we consider the problem we must first apply a little common sense. If, for example, someone asks the question *"What is the weather like?"* we should not immediately rush to create a mathematical model and a computer program in order to find the answer(s).

First we should ask "Who wants to know and for what purpose is the information needed?" If it is a member of the family who only wants to know whether to take a raincoat and umbrella, an opinion, based on just looking out of the window, may be all that is expected. If, on the other hand, the question is being asked by the pilot of an airliner due to fly five hundred people to New York, the only useful answer will involve the gathering, processing and presentation of a great many facts. In the first case a computer is neither needed nor justified; in the second case the model will be highly sophisticated and both the computer and the program(s) will be very complex.

Such examples often seem only too obvious but people do, frequently, plunge straight into doing all sorts of complicated sums and programming without hesitation. True mathematics is not like this at all; indeed it has been said that it should be "..90% think and 10% sums." Always pause and *think* about the real nature of the problem and consult the user.

2.1.4   Let us follow the process of simple modelling through the stages with an everyday example.

> The management of an engineering firm which produces aircraft parts has agreed to pay its skilled workers at an hourly rate for a basic week of up to forty hours with any time over forty hours paid at one and a half times the normal rate. Employees are only paid for completed quarter hours. The firm does not normally work during public holidays but any hours which *do* need to be worked then will be in addition to the normal forty hours and will be paid for at twice the normal rate. All employees pay six percent of their wages into a pension scheme. A payroll system is needed.

Creating such a 'narrative' description of the problem, ie a 'story' which is agreed to by the user as a correct representation, is a vital first step. However, words are often treacherous and also any verbal description will almost always include much which, although relevant to the situation and perhaps to the full final solution, is not essential to the initial model. It is here that the process of continued simplification begins.

The problem: How is each employee's wage to be decided?

(NOTE: At each stage you should discuss possible alternative ways of looking at the problem or other solutions which you or your fellow students can suggest.)

2.1.5 The stages.

Stage 1 Yes, it is numerical since it deals with time and money measurements.

Stage 2 Relevant facts:
  i)   up to 40 hours at basic rate;
  ii)   anything over 40 hours at 1½ times basic rate;
  iii) work on public holidays paid at twice the basic rate;
  iv)   final pay will have 6% deducted and paid into pension fund.

Everything else in the original 'narrative' description of the situation is now redundant so far as developing the model is concerned.Look back and see how much of it could be crossed out.

Stage 3 What we shall *need* to know, for each employee, will be:
  i)   basic pay rate (fixed, except for possible periodic reviews);
  ii)   total hours worked in any given week;
  iii) any hours worked on public holidays in that week.

This identifies a data-acquisition task which we shall, eventually, have to build into the system but for the present we can, having noted the need, continue developing the model.

Stage 4 The 'natural' stages would seem to be to calculate:
  i)   basic pay; (there may be nothing more):
        ia)   public holiday pay, if any;
        ib)   'overtime', if any;
  ii)   pension fund deduction.

Stage 5 Symbols, constants, variables.

Because stages 3 and 4 are, as they are written above, very 'wordy' we can be pretty sure that variable names and symbols *will* simplify the task.

Let us first 'declare' our variables and constants. In the following list we can, in our minds, substitute for the word 'LET', the words, "Let us agree that throughout our model . . . ". We can also interpret the '=' sign as ". .stands for. ." or "symbolises".

  LET R = the hourly pay rate;
  LET T = the total hours worked;
  LET V = the ordinary overtime hours;
        (why do I choose *not* to use 'O' for 'overtime' ?)
  LET H = the hours worked on public holidays;
  LET W= the total amount of wages to be paid to the employee:
  LET P = the amount to be deducted and credited to the pension fund.

This does not, as yet, identify the *type* of data item and we must think about that a bit further. If we assume that R implies pounds and pence, or dollars and cents, we should normally use decimal numbers but these often engender errors when converted to binary in the machine whereas whole numbers do not. Since overall amounts in this case are comparatively small, it would be wise to use whole numbers and avoid the fractions. We can do this by stating R in pence (or cents). This means declaring R as an INTEGER and, since it will be fixed, at least for quite long periods of time, we may well treat it as a CONSTANT.

T, V and H, on the other hand, will involve fractions but only 0.25, 0.5 and 0.75. (Can you see why?) These can all be converted to bicimal fractions in the machine without loss of accuracy so we can declare these VARIABLES, (since they will change from week to week) as REAL (numbers), or REALS.

We will investigate these number types more thoroughly in a later chapter but for the moment we need only accept that integers *cannot* be other than whole numbers whereas REALS may have a fraction part as well as an integer part; $2\frac{1}{2}$, 2.5, 3.63 and so on, are real numbers.

The final pay, W, will need to be set out as dollars and cents (or pounds and pence) in the end but can be calculated throughout in INTEGERS and only converted to REAL at the last possible moment to keep error to a minimum. There is, similarly, no reason why the pension contribution, P, should not be calculated as integer and converted to real at the last possible moment. Almost always there is another 'bonus' to using integers. Most computers show an appreciable gain in speed when processing integers. Why?

Again, please discuss the choices of data types and make sure that you understand what governs our choice. See also if you agree with the suggestions I have made.

2.1.6   It is at this stage that we should, in practice, begin to compile the data diction-ary in which, as part of the program documentation, we record data assignments, types and structures. We should also decide on and include in the dictionary the permissible RANGE for each data item. For example, the range for T *must* be 0 – 168. Can you see why it cannot be less than 0 or more than 168 or why V and H should be not more than 128? Try to establish all the necessary ranges because eventually we shall need them in order to stipulate test data, among other things. (Why should we test T for 'less than' 0 and 'more than' 168? One of the advantages to working as a team in computing is that we can discuss such points with our fellow members and it is important that you should get into this habit in practising model-building).

2.1.7   We are now ready to attempt to build the model itself and there may be many different ways of doing this. The 'best' model is not necessarily what we aim at in the first instance. First and foremost we need a model which *works* (is effective), continues to work *reliably* and is easy for anyone else to understand and to *maintain*. We sometimes describe such models as 'robust'.

Once we have such a model we may, if it proves slow or inefficient (or if we want to make it more user-friendly, for example), 'tweak' the model into a somewhat different form. If we make a mistake we can always go back to an original which *we know to be robust*.

For our present purpose a first step might be to state that,
      IF T≤40 THEN W = (R * T) / 100.

The '/100' element of the model will convert cents back to dollars and cents. This model is only part of what is needed since it ignores the pension plan deduction of 6% so:

$$\text{IF } T \leqslant 40 \text{ THEN } W = 94\% \text{ of } (R * T) / 100$$
$$\text{(or) } W = (((R * T)*(94/100))/100) \text{ and}$$
$$P = (((R * T)*(6/100))/100) .$$

(When we are modelling it is quite sensible to use brackets to show each step, to 'nest them' if necessary and finally to check that the number of left brackets corresponds to the number of right brackets. Most computer translators will signal a run-time error if we muddle the brackets but we can't be entirely sure – so check carefully and practise!)

2.1.8 Before we do anything else let us test this crude model, bearing in mind that we only do one step at a time and that the brackets show us the sequence in which we do them. If the brackets are 'nested', that is to say one pair within another and so on, we calculate the contents of the innermost pair first.

So, if we ASSIGN 'invented' values of 30.5 to 'T' and 250 (cents) to 'R' we have:

$$W = (((250 * 30.5)*(94/100))/100) \text{ and only the arithmetic remains to be}$$
done, in four further steps:

$$W = (((7625*94)/100)/100)$$

$$W = ((716750/100)/100)$$

$$W = (7167 \cdot 50/100) \text{ cents}$$

$$W = \$71.675$$

Now, by the same process, check that P = \$4.575 by substituting our test values of 250 and 30.5 for R and T in:

$$P = (((R * T)*(6/100))/100) .$$

If this simplified model works we can be fairly sure that we are on the right lines with the additional incentive that eventually all the drudgery of the arithmetic will be performed by the machine!

2.1.9 So far, however, we have looked at the basic week of not more than 40 hours. If the week worked were more than 40 hours a different strategy must come into play:

$$\text{IF } T > 40 \text{ THEN } V = (T - 40) - H$$
and we can fit this into the overall model:

$$W = 94\% \text{ of}((40*250)+(V*1\frac{1}{2}*250)+(H*2*250)) \text{ cents}$$

which is sound but perhaps a little clumsy. If we now apply the distributive law we can start to simplify things a little:

$$W = 94\% \text{ of}(250*(40 + (V * 1\frac{1}{2}) + (H * 2)) \text{ cents.}$$

By substituting $((T-40)-H)$ for V, (as we worked out, above):

$$W = 94\% \text{ of}(250*(40 + (((T-40)-H) * 1\frac{1}{2}) + (H * 2)) \text{ cents}$$

$$W = \$(94*250*(40+(1\frac{1}{2}*((T-40)-H))+(2*H)))/100*100 \text{ and}$$

$$P = \$(6*250*(40+(1\frac{1}{2}*((T-40)-H))+(2*H)))/100*100.$$

As always you should check the model by choosing test values for T and H and working through the steps of calculating the contents of each pair of brackets in turn.

2.1.10   If we had been solving a conventional mathematical problem which asked us to 'simplify' the equation, these models would still be capable of a good deal of refinement and we should no doubt lose marks if we left our work at this stage. Given the speed of the computer, however, we do not need to pursue matters further since the model is 'workable' and would be reasonably clear to a colleague as all the parts are still clearly identifiable.

2.1.11   What remains to be done is to connect our four partial models into a sequence so that:

$$\text{IF } T \leqslant 40 \text{ THEN } W = \$(((R * T)*(94/100))/100) \text{ and}$$
$$P = \$(((R * T)*(6/100))/100)$$
$$\text{ELSE } W = \$(94*250*(40+(1\frac{1}{2}*((T-40)-H))+(2*H)))/100*100 \text{ and}$$
$$P = \$(6*250*(40+(1\frac{1}{2}*((T-40)-H))+(2*H)))/100*100.$$

Stage 6 (and 7)   This is certainly not a set of conventional mathematics text-book models, like $A = L * B$.

2.1.12   Where else might we look? The answer, of course, would be to consider the use of a commercial PAYROLL PACKAGE in which such models would certainly be found. If, however, the user could not justify the cost of such a package we might still be asked to produce a simple program. What else is available to us?

Could a spreadsheet do the job? We should, in that case, still need our model and we might do the rest by writing suitable macros within the spreadsheet.

Stage 8   Assuming that we *do* need to write a program the final stage would be to express our model in pseudo-code form or a program structure diagram.

Throughout this book we will use the former since it is closest to a plain-language description of the task and is easily converted to any chosen high-level language when coding begins.

2.1.13 Since we need to use our program for a number of employees, we shall eventually need to embed our 'core' program segment in a loop which will deal with *all* employees but, for the moment, we need only embody our models in a simple SELECTION routine:

```
H=0; P=0; R=250; T=0; W=0;
read H, T
IF T ≤ 40 THEN
        W = (((R * T)*(94/100))/100)
        P = (((R * T)*(6/100))/100)
ELSE
        W = (94*250*(40+(1½*((T-40)-H))+(2*H)))/100*100
        P = (6*250*(40+(1½*((T-40)-H))+(2*H)))/100*100.
ENDIF
write W,P
END
```

2.2 The Algorithmic Model

2.2.1 This step-by-step process has brought us to *one* possible solution to the problem; there will be many possible variations and you should try some other ways of modelling this task.

What we have arrived at is a straightforward algorithm. If we input the right values for T and H (and R is kept up-to-date) we must arrive at the right answers.

2.2.2 An example of a similar simple commercial model which we might meet is:

A finance company charges 15% per year on all loans. A program is needed to calculate the total amount to be repaid after any given number of years.

We can, this time, abbreviate some of our steps:

Stage 1   YES
Stage 2   15% per year.
Stage 3   Amount borrowed and number of years the loan 'runs'.
Stage 4   YES, a new calculation for each year.
Stage 5   A = amount borrowed (variable); D = the amount due to be repaid (variable); R = rate of interest as % (constant); T = the time for which the money has been borrowed, in years (variable).
   (Discuss with your colleagues the best numeric data *types* to use).
Stage 6   YES – substituting our chosen symbols for the traditional mathematical ones in the conventional 'compound interest' formula:

$$D=A(1+R)^T$$

Stage 7 Probably not. Most people would not immediately recognise what this formula does and that is a very good reason for not using it in a program.

It will be simpler, clearer and easier to test and maintain if we set up a more straightforward model which will repeat the calculation, by a loop, for each year. The sheer speed of the computer makes such repetitions a practicable alternative to the one-time algorithm. Let us try to create such an 'iterative' (ie repetitive) model.

At the end of the first year the borrower will owe:

The amount borrowed (A) + 15% of that amount (R*A) = D

At the end of the second year the borrower will owe:

D + 15% of D = D (the new value of D)

– and so on, for the number of years the loan runs.

When we use a value arrived at in the previous loop in order to calculate the current 'pass' we call this a RECURSIVE iteration (and a very powerful tool it can be).

In the present case we have a fairly simple model:

D = D + ((15 * D) / 100)

and if we start with inputting values for the two variables, A and T, we have a complete model.

2.2.3 *Stage 8* Pseudo-code:

```
        A=0; D=0; R=15; T=0.
        read A,T
        D ← A
                DO UNTIL T=0
                D ← D+((R*D)/100)
                T ← T−1
                END DO
        write A, D
        END
```

This model is capable of being developed for many commercial tasks. It could, for example, be converted to monthly calculations for, say, a credit card company. Once again you should experiment with the model, adapt it to other purposes, test it by 'dry running' and so on.

2.2.4 Discuss other simple commercial or industrial problems with your fellow students and systematically work through the process of modelling them. Don't worry if you and your colleagues come to different conclusions, there will almost always be a number of possible solutions to any problem. If it works, when you dry-run it, your solution is likely to be practicable. It is also important to remember that almost any solution is capable of refinement and improvement; the real test is whether your model is well thought-out and reliable.

Do not ever be content simply to be told, "This is the answer". You may well develop better, more efficient answers by using your own experience and your own ideas.

2.3 A Model of a Game

2.3.1 Let us, in order to build up some versatility, look at a different kind of problem.

A popular English table game is known as Snakes and Ladders. On a very brightly coloured board, of one hundred squares numbered in sequence, are printed several snakes, and a number, not necessarily the same, of ladders of different lengths. See Figure 2.1. The players move their counters around the board in turn, according to the roll of the dice. As each moves along the squares he or she may land on a square containing the bottom of a ladder or the head of a snake. In the former case the player moves to the square at the top of the ladder and *gains* ground: in the latter case the player is 'swallowed' by the snake and slides right down to its tail. The first player to reach the 100 square wins the game. It is a simple game of chance but can be great fun with plenty of 'highs' and 'lows' along the way.

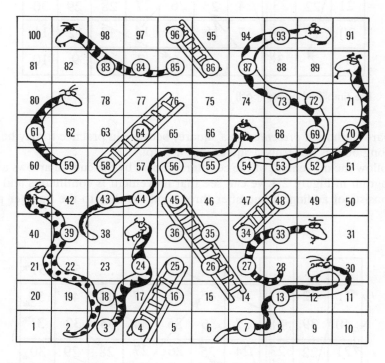

Figure 2.1 A black and white, simplified, Snakes and Ladders board

2.3.2 Many people, faced with writing a computer program for a Snakes and Ladders game, would be preoccupied with the graphics, shown in Figure 2.1. It is especially important that we should stop and think!

As always, let us simplify the model. This means eliminating all unnecessary detail (so far as the model is concerned).

2.3.3 First of all the snakes and ladders may be represented by simple arrows and we may substitute lines going from one position to another, (or VECTORS), at this stage even though, much later in the programming sequence we may wish to restore full colour graphics to the screen! See Figure 2.2.

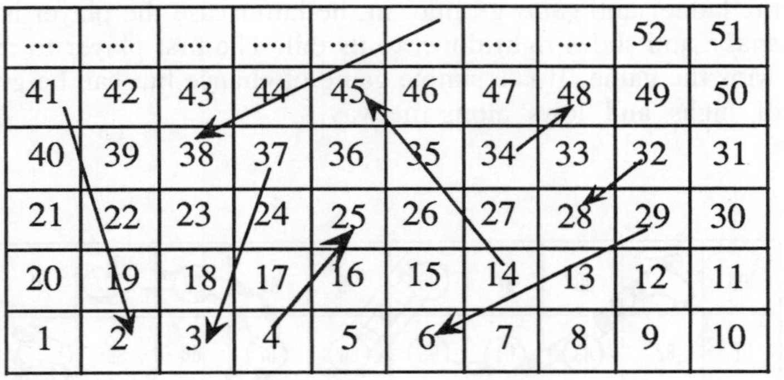

Figure 2.2 The simplified board (part)

The board, although *presented* as a square, because in practice it may be folded to fit an oblong box, is not truly a square but a strip!

If we follow the movement of the counters, shown by the curved arrows at the sides of the diagram in Figure 2.3 we can see that movement is continuous and that each 'move' consists of adding the number thrown on the dice to the number previously reached.

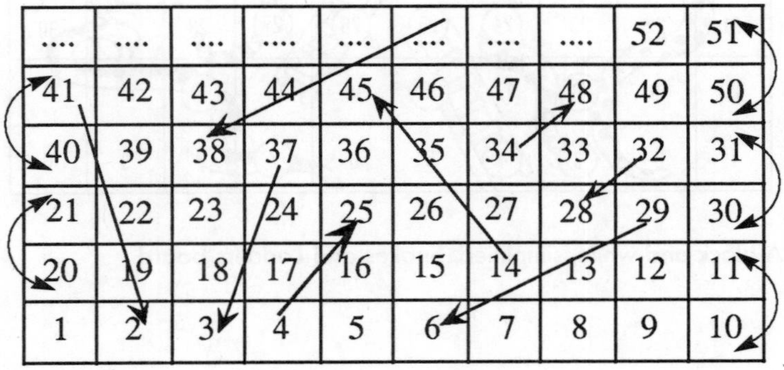

Figure 2.3 The pattern of movement

2.3.4 A simplified diagram which puts together the arrows, as the abstract version of the snakes and ladders, and the strip, instead of the square, might look something like the one in Figure 2.4.

THE NUMBERED SQUARES AS A STRIP

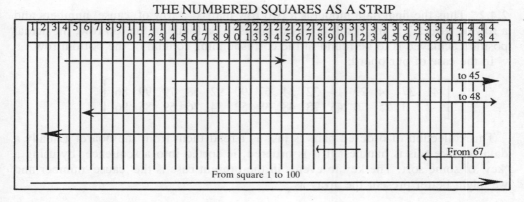

Figure 2.4 The simplified *visual* picture

2.3.5 This process of repeated 'abstraction', that is to say, moving away from reality to a chosen *model* of reality, is similar to that employed by artists who tempt our imaginations with something other than a photographic likeness of the world. Even the photograph is, however, only a model of the reality which we see. At best it allows us to store and communicate to others our view of that reality. The mathematics we are discussing closely resembles art in many ways.

2.3.6 The last stage in simplifying this model is to make the journey from the visual model to one which is purely numeric. Even the arrows and the strip are not absolutely essential since we can say, of any one 'move', "If I reach 100 I have won." Otherwise, at the simplest, the new P(osition), often indicated as P′ will equal the old position, P, plus the throw of the dice. This is fully represented by:

$$P' = P + T(hrow)$$

2.3.7 Only if P′ is at the foot of a ladder or the head of a snake is there any complication, a 'slide' from that square to another. All such slides can be described by, "If my counter lands on a *cause* number (or square) it ends up on an *effect* number (or square)." We already have a fine mathematical tool for holding pairs of numbers like these. We call it an 'ordered pair' and the relationship between any such pair of numbers is preserved by our always dealing with them in the same sequence, in this case CAUSE followed by EFFECT.

2.4   The Matrix Model

2.4.1 When we have a number of such ordered pairs the best way of keeping them all in place is to put them into a *table* or an *array* or, as it has known to mathematicians for generations, a *matrix*, (*matrices* in the plural). (There is no essential difference between any of these names and in computing we have simply fallen into the habit of making rather more use of 'table' and 'array'.)

2.4.2   Thus the final data model is a two-row matrix; a row of 'cause' numbers and a row of 'effects'. If the result of any throw is to be found in the top row, the actual position achieved is to be found in the corresponding column of the bottom row.

In the case of our board:

$$\begin{bmatrix} 41 & 37 & 4 & 29 & 14 & 32 & 34 & 67 & 90 & 92 & 58 & 79 & 99 & 86 \\ 2 & 3 & 25 & 6 & 45 & 28 & 48 & 38 & 52 & 54 & 76 & 59 & 85 & 96 \end{bmatrix}$$

This collection of ordered pairs is just as I read them from the board. Usually a data set like this is handiest if it is ordered, ie sorted into a sequence, by using a 'key' of some sort. The key to events in the present case is the 'cause' number and we might better present the array as:

$$\begin{bmatrix} 4 & 14 & 29 & 32 & 34 & 37 & 41 & 58 & 67 & 79 & 86 & 90 & 92 & 99 \\ 25 & 45 & 6 & 28 & 48 & 3 & 2 & 76 & 38 & 59 & 96 & 52 & 54 & 85 \end{bmatrix}$$

and we can then examine it, after each throw, in sequence.

2.4.3   We can think of a matrix as a number of rows and columns of data-holders, or 'cells'. As is so often the case with mathematics we turn to the shorthand of symbols rather than using lengthy descriptions. For matrices we designate the matrix by a capital letter, let us call ours 'C', and each of the data elements stored in the matrix is identified by giving its position by its row and column address. We designate these by using a small 'c' for the element and small numerals set at the foot of the 'c' on the right-hand side, *subscripts*, to indicate, as an ordered pair, first the row number and then the column number of the *address* of the *cell*.

2.4.4   Because it is accepted practice to use 'i' as a variable name to denote the row and 'j' to denote the column of an address the general form of cell address is given by: $c_{i,j}$

In the present case, the last ordered pair in our sorted matrix, 99, 85 would be designated as $c_{1,14}$ and $c_{2,14}$ respectively and the fourth pair, for example, would be $c_{1,4}$ and $c_{2,4}$.

This notation may be familiar to you but if it is not you should make sure, by practising on our model, that you can use it freely because it is of real importance.

2.4.5   The final stage is to express our model as the core program segment, ie, making one move for one player. We shall eventually need to allow for more than one player and if we set up a smaller array with 'n' cells for 'n' players then each player's position on the board, initialised to zero, will be designated by P_1, P_2 and so on.

Working out the model for one player only, using a pseudo-random number generator instead of a die, we have something like:

```
DO WHILE more moves are to be made
j ← 1, T ← RND(6), P₁ ← P₁ + T
    IF P₁ = 100
        display 'You have WON', END
    ENDIF
        DO WHILE j < 15
            IF c₁,ⱼ = P₁ THEN P₁ ← c₂,ⱼ
            ELSE j ←j+1
            END IF
        END DO
    display P₁
END DO
```

Once we are sure that this, or something similar, actually works we are ready to begin working 'outwards' from the core segment to add all the other embellishments, including the beautifully coloured graphics, which will bring the game to life on the computer screen.

2.5 A Numerical Model

2.5.1 So often, in practice, we find that quite complicated situations may be reduced to comparatively simple numerical models.

2.5.2 A very early mathematical model was developed by the great Italian mathematician Leonardo Fibonacci, or Leonardo da Pisa, some seven hundred years ago. He was modelling a scientific study of how some animals multiply. We will follow his modelling process.

> Let us assume that a female rabbit comes to breeding maturity and has her first litter at two months of age. She is able to produce a further litter every month thereafter. If the worst case* assumption is that only one female from each litter survives to breed how many female rabbits will this generate in one year?
> (* Scientists often assume the 'worst case', ie what is the most pessimistic view of a situation.)

2.5.3 In the first instance we might illustrate this situation by a traditional family 'tree'. TREE models show changes, developments, or generations on each successive level, i.e as trunk, bough, branch, branchlet, twig, leaf, with as many 'branchings' as the model needs.

If we illustrate our rabbits by drawings, which are themselves, of course, models or abstractions, we find the tree developing as shown in Figure 2.5.

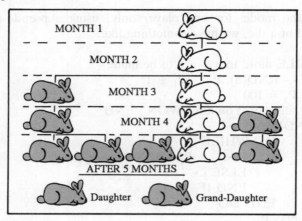

Figure 2.5 The rabbit tree

This, of course, would become very elaborate after a very few months and we clearly need something less time-consuming to draw and which takes up less space.

2.5.4 The remedy varies from one application to another but we usually represent each 'branch' by a pair of lines which 'grow apart' from each other at an angle.

2.5.5 For the present purpose I have chosen to use a single line (segment) to represent our 'standard unit' of one female rabbit.

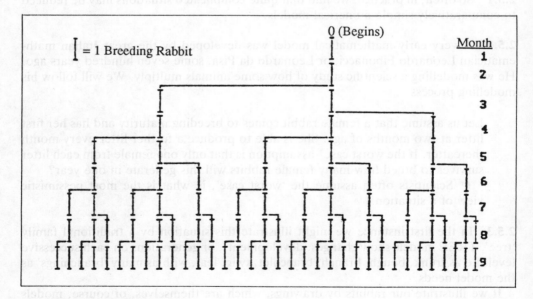

Figure 2.6 The 'abstract' tree

2.6 A Tree Model

2.6.1 The resulting Figure, 2.6, is a tree diagram which has been converted, almost completely, into an abstract visual representation of our information. Such tree models have many uses and we shall examine them more thoroughly later in the book.

In the present case any pattern which there might be is rather obscure but we *can* see that growth is at an increasing rate.

2.6.2 Fortunately this is another situation where we can go further and extract a numeric model which may clarify matters. We first look at the month-by-month record and *count* all the rabbits in our diagram. Whenever we reach this stage in the development of a numeric model we shall usually find it profitable to TABULATE the data, ie put them into a table. The next step, in looking for any patterns which may help us, is to look at the differences between successive pairs of numbers, in this case the size of each month's increase in the number of rabbits.

2.7 A Table Model

2.7.1 Setting these results into Table 2.1 we see that:

Table 2.1

Month	Start	1	2	3	4	5	6	7	8	9
Rabbits	0	1	1	2	3	5	8	13	21	34
Increase			0	1	1	2	3	5	8	13

2.7.2 Can you complete the table for the twelve months?

2.8 A Model as a Sequence

2.8.1 This pattern of growth in the numbers of rabbits is, when we study it, completely predictable and any such set of numbers is known as a SEQUENCE. This is another of those words which has a special meaning in mathematics. It means not only that the numbers, or 'terms' as they are often called, follow one another but that we can always say just what the next number in the sequence will be. This particular model actually describes many thousands of natural growth patterns and is renowned as the 'Fibonacci sequence', after its great discoverer.

2.8.2   A program to model and develop this sequence is fairly straightforward:

$S_1 = 0$, $S_2 = 1$, $i = 3$
Accept T; [number of terms needed]
  Do until i = T
    $S_i \leftarrow S_{i-2} + S_{i-1}$
    write S_i
    $i \leftarrow i + 1$
  END DO
END

2.9   Conclusion

The process of modelling requires a knowledge of a range of mathematical techniques but is, nevertheless, one in which we can become competent by following a consistent routine. For the computer the most elegant model is not the first requirement. Developing one which is true, robust and which *works* is our first concern; the elegance can come later as we become more experienced and develop a pride in the *quality* of our models.

Exercises

1) Develop a pseudo-code model of a stock valuation procedure which will include, for each item in the stock-list, RECEIPTS, ISSUES and STOCK-HOLDING, day-by-day, with both the quantities *and* prices of receipts and issues, varying from day to day.

2) Expand the model of the Snakes and Ladders game to allow for up to *six*, *named players*.

3) The sequence in 2.8 produces an interesting variation when it generates the set F in which $f_n = f_{n-2} / f_{n-1}$.
Develop and dry-run a pseudo-code model of this.

4) Develop a model to calculate lorry transport costs per km. if, for each vehicle:

   the distances run each month are highly variable;
   fixed costs are $700 per month;
   variable costs are $2 per km;
   depreciation is $500 per month.

5) A new payroll tax has been introduced. It is to be collected by the employer and paid by him to the Government. Employees and employer contribute jointly with the former paying 1% of gross earnings per week or per month depending on how frequently they are paid. The employer must add 1½ times the amount contributed by the employee.

   a) State this as a mathematical model;
   b) write a pseduo-code routine to be added to the current payroll program in due course;
   c) devise appropriate test data and dry-run your routine. Debug if necessary.

6) Model, in pseudo-code, a range-check routine to check the total of every tenth employee's pay to ensure that it is not a negative amount and also that it does not exceed $1000. Each full payroll run of this routine should start with a different member of the set {the first ten employees on the payroll} chosen at random.

7) a) Develop to a *proven* pseudo-code segment a model which will calculate and print a *two-way* timetable for local trains starting at 'A' and 'E' respectively based on the following data.

The trains are to depart at 20 minute intervals from 7 o'clock in the morning to 7 o'clock in the evening. They run at an average speed of 40 kph and the distances covered are: A to B – 10 km; B to C – 8 km; C to D – 14 km; D to E – 16 km. The timings must allow for station stops of 2 minutes at B and C and at least 8 minutes at A and E.

b) What is the minimum number of trains needed to run the service if one is always to be kept in reserve?

3
Sets and Their Applications

Objectives

By the end of this chapter you should be able to:
- understand the logical requirements of a set;
- define, enumerate and specify sets;
- solve simple problems using sets;
- understand the relations between sets;
- derive simple mappings and functions.

A REMINDER: In the Introduction to this book we saw that to be a 'proper' mathematical set a collection must be DEFINED and EACH OF ITS ELEMENTS MUST BE UNIQUE.

(If, in practice, we have a collection which is *defined* but includes items which *are* duplicated we sometimes refer to it as a 'bag' and it cannot be dealt with by the rules of sets.)

3.1   Defining a Set

3.1.1   When discussing a payroll model in the previous chapter we looked briefly at the test data which would be needed to validate H, the hours worked. We saw that both H and the rate of pay could, for the sake of improving processing accuracy, be expressed as INTEGERS. In the case of H we did not make a suitable conversion at the time; let us now do so and consider the implications for valid inputs to H and for the essential test data.

The hours were to be recorded *only in completed quarters* so that someone who worked, say, for 7 hours and 18 minutes would be credited with 7¼ hours only.

In order to use integers we should convert all values for H to the equivalent number of *quarters* and, at the very end of our calculations, divide the calculated pay by four to arrive at a final total. Our 7¼ hours equals 29 quarters.

3.1.2   For practical programming purposes we must now examine the implications of this. In one week there are 7*24 hours, 168 hours in all or *672 quarters*. No-one can possibly work more than this. Similarly no-one can work a *minus quantity* of hours! (Although this sounds absurd it is entirely likely that someone, by simple human mistake, would sooner or later enter a negative amount and we must therefore expect it and guard against it.) Our data range is, therefore, from 0 to 672, inclusive. In mathematical symbols we should write: $0 \leqslant H \leqslant 672$.

Since such expressions may not be immediately familiar to all of us let us translate it back into words: "0 is less than or equal to H, the number of quarter hours worked, which is less than or equal to 672.". This is still clumsy and may be better understood as, "H must be at least 0 and at most 672."(*)

3.1.3 One of the most difficult things, for any programmer (or anyone reading a program listing), is to ensure that any CONDITIONS which control a program are LOGICALLY CORRECT. This is precisely where we can best use the language and principles of SETS to help us clarify our thinking. Note that the proper time to turn to this branch of mathematics is *before* we start writing code, ie, in the *design of our program*, as we seek to develop the core model into a practical system.

In the present case, by defining our data range as $0 \leqslant H \leqslant 672$ we are, without stating it *explicitly*, also declaring that *any other value is wholly unacceptable*. This is an aspect of logical expression which is not common in everyday discourse (discussion or exchange of ideas), and it is quite difficult to learn to think in such a rigorous way.

Unfortunately the computer is quite unable to tell right from wrong, logical from illogical, and we *must* learn to think like this if our programs are not to be riddled with errors of logic, as is all too common. Put quite simply; every time that we define a logical IS we also define a corresponding IS NOT. What, then, *is not* within the acceptable data range for H? We can state the conditions for both H and NOT H in many ways and making elaborate lists is one way.

3.1.4 The simplest way is to define H as a SET and we can do this in several ways:

a) *by a verbal description such as that shown in 3.1.2 by the (*)*; it is always possible to describe such situations fully in words if one is not comfortable with mathematical symbols but where logic is concerned, if we are to avoid any possible ambiguity or misunderstanding, this can be a very long-winded process.

b) *by listing all the elements (members of the set)*; this is fine for small sets but becomes both cumbersome and prone to mistakes for sets of any but the smallest size. See Figure 3.1.

c) *by abstract mathematical symbols*; this is, without doubt, the most compact description of a set. We will look carefully at this symbolism presently and investigate all the symbols which we use in order to learn this most valuable shorthand as we go along.

d) *by Venn diagram*; this for most people is clear and easy to work with yet provides us with a very powerful tool, as we shall see. Its limitation is that it is often difficult to display large numbers of elements except as quantities.

Figure 3.1 The Venn diagram can get crowded

3.1.5 What is our 'universe of discourse' (usually given the mathematical symbol &) in this case? The only valid data for our model is the *number* (of quarter hours) assigned to H. The variable is *numeric* and we should usually see a run-time error message if we tried to input anything but numeric data. (Nevertheless we cannot take this for granted and must check the particular compiler or interpreter which we shall be using to ensure that it does.)

3.1.6 In this case our &, the universal set, is the *whole world* of numbers; of which we are prepared to accept as valid only *0* or *positive whole numbers up to and including 672*. See Figure 3.2.

What must be rejected is all which is *outside* this small set of valid data.

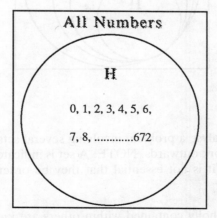

Figure 3.2 Our small data set in a very large 'universe'

3.1.7   A Venn diagram will help us see the whole picture. What is *invalid* is *all* other types of number including integers <0 and all real numbers which are not positive integers ≤672. Since our test data must always include examples which should *not* be accepted, as well as those which will show that the program does run successfully, we must select sample values from ℰ also.

3.1.8   The entire area of the universal set outside a named set is known as the COMPLEMENT of that set and is designated by a ' symbol. Our ℰ consists of H together with H′.

3.2   Relations between Sets

3.2.1   The small part of ℰ which *is* acceptable to us is, however, part of several other sets of numbers and the less confident we are with traditional mathematics the more valuable we are likely to find it useful to detail H′ more carefully as in Figure 3.3. Here we begin to clarify the complex relationships within ℰ and identify our valid and *invalid* data more readily. Remember that everything *outside* any set is 'NOT', in our case *not* valid.

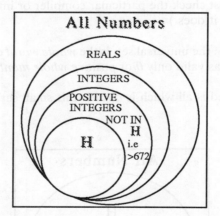

Figure 3.3   Sets within sets

3.2.2   Whenever we analyse a problem involving several sets it is best to start from the innermost set and work outwards. NOTE: A set is indicated by placing the name, or the list of elements (it is not essential that they be ordered), between 'braces', {1, 2, 3, 4}.

3.2.3   Sets which are wholly contained within others are known as SUBSETS and designated by the special symbol, ⊂.

3.2.4   Our current diagram tells us that:

H ⊂ P ({Positive Integers}. But all of the latter which are >672 are INVALID for our task.)
P ⊂ I ({Integers}. But only a subset of +ve integers is valid for us.)
I ⊂ R ({Real Numbers}. Any of which having fractional parts are invalid for us.)
R ⊂ & (All Numbers).

3.2.5   We now have a logical model which indicates what data-testing we need to do in our validation procedures for H.

(NOTE: the *new symbol* ≠ signifies 'does not equal' or 'is not'. The programming language reserved word, INT(n), which we will also use in our pseudo-code, signifies 'take the whole number part only of 'n'.)

```
        IF H ≠ Numeric then INVALID
            ELSE IF H − INT(H) > 0 then INVALID
                ELSE IF H < 0 then INVALID
                    ELSE IF H > 672 then INVALID
        ELSE H is VALID
        ENDIF
```

Make sure that you can see what each line is meant to do and, as always, check very carefully to make sure that the segment works and that it accurately represents the logical structure of sets which we have been developing.

3.2.6   This is a good point at which to use our example to introduce some other new symbols so that we may learn to use them without having to write verbal descriptions. That, after all, is their main value but equally important is the fact that the symbols mean the same the whole world over whilst words have to be translated from language to language!

3.2.7   To clarify still further what we discussed earlier, a set, no matter how large or small, may be designated by NAMING it, for example {the set of integers}, or by a LETTER, in our example, H.

We may LIST the elements of a set, say A = {cat, dog, horse, wolf, tiger, panther, lion, elephant}, if it contains only a few ELEMENTS. We could then say, for example, 'tiger ∈ A'. (Tiger is an *element* of the set which we have labelled 'A'.)

Note very carefully that we are not implying that the set 'A' designates any particular ATTRIBUTE of its elements; the set is *not*, for example, *necessarily* a set of mammals although it would be, without a doubt, correct to say, if the images in our illustration of this set in Figure 3.1, represented real animals, as opposed to, say, furry toys, that 'A ⊂ {mammals}'.

You should check and discuss all such statements to make sure that you and your fellow students follow the *logic* of the situation and are learning not to jump to conclusions.

3.2.8   Any small set, such as A, can also easily be ENUMERATED or NUMB-ERED, ie counted. We often use 'n' to stand for any number, known or unknown, in mathematics and we can say that n(A) = 8 or that A has 8 elements. Very large sets may be impossible to number even if they are FINITE, ie have a fixed number of elements whilst many sets are INFINITE. Discuss what this means and decide what effect the *size* of a set, of a file perhaps, may have in practical computing terms.

3.2.9   Let us, for its designation in ABSTRACT SYMBOLS, return to our example. Our set, H, can be designated as:

{0 together with the +ve integers ≤ 672}.

This is still rather clumsy and we are better writing
{0, 1, 2, 3, ...672}(† see below).

The more purely symbolic form you may find written in mathematics textbooks as:

$$H = \{0; h_i : 1 \leq h_i \leq 672, h_i \text{ integer}\} - \text{or something similar.}$$

Do not be intimidated by such statements! We can retranslate them back into plain English a step at a time:

'H designates a set which consists of zero and individual elements, h_i, which are at least 1 and not more than 672 and all of which are integers, or whole numbers'.

There are many occasions when it is useful to abbreviate our descriptions in this way but if in doubt, use the 'listing' method, shown by †, above.

3.2.10   There are a number of other symbols which we need to use and this is perhaps a good time to list them:

=	Is equivalent to
≈	Is approximately the same as
≡	Is the same in all respects as
&	The Universal set, or Universe of Discourse
{....}	The set
∈	The item to the left is an element of the set to the right (of the symbol)
∉	Is not an element of
⊂	Is a true subset of
⊃	Has, as one of its subsets,
∪	The union of sets (eg A ∪ B) puts together all the elements of A with all those of B
∩	The intersection between (eg A ∩ B) groups all elements which are in *both A and B*
/	This sign makes any other symbol into 'NOT'
∞	Infinity – 'goes on for ever'

3.3 More Complex Relations

3.3.1 There are many ways in which we can represent sets by Venn diagrams and we should pause to look at a few typical examples. Some of the relationships are indicated in Figure 3.4. See if you can fully explain the logical story told by examples a) to h), in words and, if possible, in symbols.

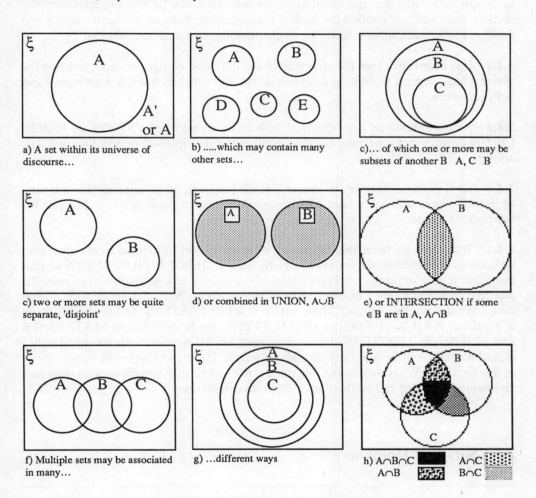

a) A set within its universe of discourse...

b) .....which may contain many other sets...

c)... of which one or more may be subsets of another B A, C B

c) two or more sets may be quite separate, 'disjoint'

d) or combined in UNION, A∪B

e) or INTERSECTION if some ∈ B are in A, A∩B

f) Multiple sets may be associated in many...

g) ...different ways

h) A∩B∩C   A∩C
   A∩B   B∩C

Figure 3.4 There are many possible variations on the basic Venn diagram

3.3.2 You should, with your colleagues, select various 'universes of discourse' and explore the sets within them. For example, the universe of four-sided, straight-line, mathematical shapes which includes rectangles, squares, oblongs, parallelograms, trapezia, rhombuses, irregular shapes .... and perhaps others!
Can you see why we need the word 'oblong', which many people wrongly think of as slang?

3.4   Practical Set Models

3.4.1   An extremely important application of sets, in computing, is that governing an organisation's DATABASE, that is to say the total storage system for all the information needed to run the organisation.

This vast universe of discourse will contain three major subsets: the first, the information needed for the day-to-day routine activities of the organisation; the second, that which is needed for middle-management, tactical, decision-making and finally, that needed for top-level, strategic planning and decision-making.

3.4.2   It is imperative that the structure of the database be rigorously logical and the systems analyst must be able to establish all the sub-sets and the relations within and between them.

3.4.3   As an instance of how involved such relations may be, a STOCK or PARTS LIST is shown as an especially complex example of multiple sub-sets in Figure 3.4g), above.

3.4.4   As a practical task try to create a stock list, with all its appropriate sub-sets, of the stationery and equipment which is needed by your present study class just for the present lesson.

3.4.5   What follows from this taxonomy of stock is that individual items must also be named, consistently, according to strictly logical DESCRIPTION RULES so that everyone in the organisation knows into which subset any particular item must be placed. Once again, try a number of examples, such as 'Nail, flat-head, wire, 75mm long.' What does this tell us? That, within a set of FASTENING DEVICES there is a subset, NAILS, of different HEAD TYPES, made of different MATERIALS and of different LENGTHS. There may well be flat-headed nails made of other material; there will be a variety of nails of length 75mm, so there will be a number of intersections. What we need at this stage is to explore such situations with diagrams and *careful thinking* rather than any elaborate calculations.

3.5   Sets and Domains

3.5.1   When we look at a Venn diagram we see that the boundary lines of the different sets create *regions*, which we generally refer to as DOMAINS, within the universal set.

In Figure 3.5 for example there are 8 DOMAINS in all; let us identify them clearly and then label them in formal symbols. (For the time being we will continue to use the '−' symbol for 'less' or 'excluding' but we will use the '∪' symbol for combining sets. NOTE: it is always easier to read such sequences of symbols if we keep the letters in alphabetical order.)

The relevant descriptions are:

1 (A∪B∪C)'
2 A–((A∪B)∪(A∩C)) or, in words, "A less what is in A but is *also* in B and C."
3 A∩B
4 B–((A∩B)∪(B∩C))
5 A∩B∩C
6 A∩C
7 B∩C
8 C–((A∩C)∪(B∩C)).

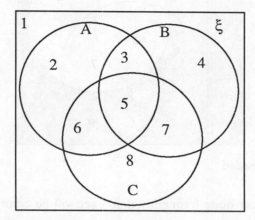

Figure 3.5 The domains identified

3.5.2 We shall need, for more precise labelling, two other symbolic representations.

When we talk of the set A we describe what is IN A. But we also imply that the elements in any other region or regions, however complex, are NOT IN A, as shown in Figure 3.6.

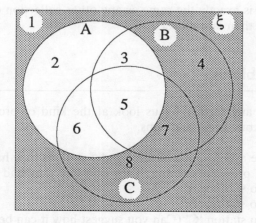

Figure 3.6 The shaded region is 'NOT A'

3.5.3 This, in the case of a single set within its universe, we have already seen to be the *complement of A* or A' or whatever is needed to *complete the whole* (universe of discourse). However we often need to identify, as NOT(IN)A, one or more of the elements in other domains. This may be (much) less than the complement of A and we use the representation $\overline{A}$ or, for example, $\overline{A \cap B}$ means "what is NOT in the intersection of A with B".

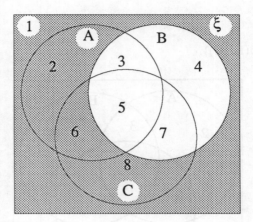

Figure 3.7   NOT B is shaded

3.5.4 It will also occur quite frequently that a set will be empty, ie it will have no elements.

We could, for example, quite logically define {human beings over 3m tall} and find, within any particular population that the tallest person is 2m in height.

Similarly there will be sets within a database which, from time to time, are empty. Most shopkeepers will hope that the set of 'bad debtors' will *normally* be empty but the set must still exist within the database.

3.5.5 We call such sets 'null' with the symbol ∅. In rigorous logic, since any collection of nothing is exactly the same as any other collection of nothing we speak of *the* null set rather than 'a' null set.

3.6   Solving Problems

3.6.1 Before going any further let us look at the kind of problem which can be solved using nothing but Venn diagrams.

100 people were asked which of three radio stations they had listened to at any time during the previous evening. They said that they had listened as follows:

65 to station 'A'

45 to station 'B'

55 to station 'C' (Can you suggest how it can be that these numbers total far more than a hundred?)

Several people had listened to more than one station at some time during the evening:

22 to *both* A and B

32 to *both* B and C

35 to A and C and 12 people had listened to all three. What story do the 'listening' figures actually tell us and were there any who didn't listen to any of the three stations?

3.6.2 When we see a situation like this, including lots of numbers, most of us are tempted to pick up our pencils and start doing 'sums'. As always, let us look and think a bit before we start doing anything.

Since the numbers we are given actually add up to far more than 100 ordinary arithmetic might be a bit confusing! A further look shows that we are dealing with a total of 100 people who fall into three 'categories' or into combinations of those three categories.

3.6.3 Whenever we are 'classifying' data we should first look to see if those groupings are or can be sets. What we have in this case is the set of 'people who listened to station A' and similar sets for B and C. We can now model this problem, or any others like it, by a Venn diagram.

LET A={people who listened to station A}
LET B={people who listened to station B}
LET C={people who listened to station C}

and since there are some people in each of the possible combinations of A, B and C we need a 'three-set intersection' diagram into which we can put our data. Figure 3.8 shows us what it looks like.

The letters denote the sets and the numbers in heavy type the domains, just as in our earlier example, Figure 3.5.

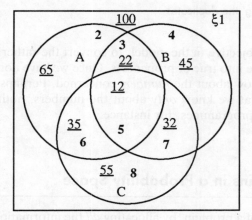

Figure 3.8 The 'raw' numbers allocated

3.6.4   Our actual data are inserted as they were originally listed but we have not yet adjusted them to total 100.

3.6.5   Here is where we start from the centre and work outwards. The people in region 5 are also in regions 3, 6 and 7. Can you see why?

Those people who listened to A, B *and* C are *also* included in the numbers of those who listened to either A and B; A and C, or B and C. Similarly the number of people reported as listening to A includes those who listened *also* to B, *or* to C, *or* to *both* B *and* C.

The rules of sets, however, dictate that there may not be duplicates of any element in a set so those in domain 5 must be subtracted from each of domains 3, 6 and 7.

In exactly the same way the 10+12+23 people already represented in domains 3, 5 and 7 respectively are all in A and their total must be subtracted from the total number of elements in A, or n(A).

So 65-(10+12+23) leaves 20 people *who are in A ONLY.* Finally we see that 'C only' is NULL and that 12 people did not listen to any of the three stations. This can be seen clearly in Figure 3.9.

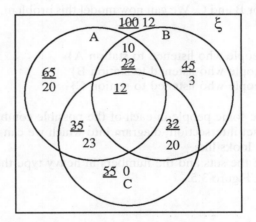

Figure 3.9   'Cancelling' the duplicates

3.6.6   This is our objective in the model, to convert the rather muddled information given in the narrative into true or proper sets. Once we have done this there is nothing which we do not know about the *numbers* concerned. Perhaps we should, however, remind ourselves that we know *only* about the numbers, nothing about the *quality* of any of the radio programmes, for instance.

3.7   The Domains in a Probability Space

3.7.1   This model of a problem, by allocating *all* the information to separate domains and refining it, actually gives us another very useful tool.

3.7.2 We have created, almost as a by-product, a 'map' of probabilities or a 'probability space', bounded or limited by &. We shall be exploring the ideas of probability much more fully in a later chapter but let us just glance at the story our 'map' tells us. For example:

If we met one of the 100 people represented by our & what is the likelihood that he or she would have listened to:
a)   station A;
b)   both A and C;
c)   none of the three stations;
d)   at least two stations ?

3.7.3   a) 20 of the 100 people listened to A only so the likelihood of our one person having done so is 'twenty out of one hundred', 20/100 or 'one in five'.

b) All the people in regions 2, 3, 5, 6, 7 and 8 listened to either A *or* C but only those in regions 5 and 6 listened to *both*. So there are 35/100 chances of our selected person having done so (or 7/20). Look at Figure 3.10.

3.7.4   You will, no doubt, now be able to answer c) and d) for yourself. Try also to label and quantify several other probabilities from our map; there are many to choose from.

3.7.5   Such probability models are rather easier to work with than many others and are well worth practising.

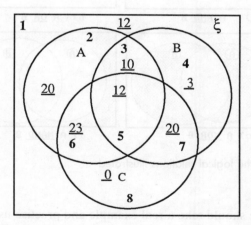

Figure 3.10   A probability 'map'

3.8   Other Logical Relations

3.8.1   Before we leave our Venn diagram models we should look at one other aspect of notation which we shall be using extensively in a later chapter.

3.8.2   For purely symbolic logic descriptions and operations, as well as in programming logical decision-making statements, we use the words AND, OR, XOR and NAND to describe relations between sets. The following illustration, Figure 3.11, shows what they convey.

a)   the shaded area shows the domain which contains all those elements which are in *both* A *and* B;

b)   the shading identifies everything which is *either* A *or* B;

c)   here we have what is in A *or* B *but not in both*;

d)   indicates what is *not in* A *and not in* B.

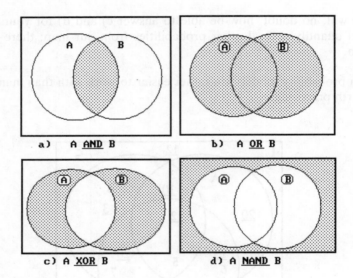

Figure 3.11   Four of the logical relations illustrated

3.8.3   As always we should take a real example and practise using this notation. We will often find it valuable to test programming statements such as:

$$\text{DOWHILE } A < B \text{ or } A < C \text{ and } A > D$$

by modelling them as a Venn diagram to make quite sure that our statement actually does what we want it to do. Look at this one carefully; does it, for instance, need any brackets to clarify its purpose?

3.9 Mapping

3.9.1 There is an extensive range of relations between sets which are between the individual *elements* of two or more sets rather than between their wholes. These are best explored using ordered listings rather than diagrams.

3.9.2 For example there is a clear relationship between: {cat, dog, horse} and {foal, kitten, pup} but we may see the relationship better if we order our data and set out the model as:

P = {cat, dog, horse}
O = {kitten, pup, foal}

when each element in the 'top' set may be linked visually to its corresponding element in the 'lower set' and we can clearly see that the relationship is that of 'parent to offspring'.

3.9.3 Such a correspondence between the elements of sets is called a 'mapping' and the relation between such sets is known as a 'function'.

3.10 Functions

3.10.1 The function, in this example, might be described as 'parenthood', or something of the sort.

3.10.2 In another example:

S = {cat, home, ship, banana}
P = {cats, homes, ships, bananas}

which is a mapping of the 'string' elements of S on to P by the function "P_i" = "S_i" + "s".

Translating back into plain language we might say, "From any string in the set, S, we may derive the corresponding string element of P by adding, (appending or concatenating), the string "s" ."

In computing we often have to deal with text and string functions so that such concatenations of strings are routine operations.

3.10.3 For the moment it may be a little easier to look at some number sets and their mappings and functions.

3.10.4 { 0, 1, 2, 3, 4, 5 . . . . ∞}
↓, ↓, ↓, ↓, ↓, ↓,
{ 0, 2, 4, 6, 8, 10 . . . . ∞}

This mapping shows, in very formal symbolism, a situation with which we have all been familiar since we were quite young. What is the function?

3.10.5   This shows a ONE-TO-ONE CORRESPONDENCE, in other words for each element of the 'parent' set there is a corresponding element of the 'derived' set.
Here is another kind of relation:

$$\{1, \quad 1, \quad 2, \quad 3, \quad 5, \quad 8 \quad 13 \quad \ldots\ldots\ldots \infty\}$$
$$\{ \ 1, \quad \cdot5, \quad \cdot66, \quad \cdot6, \quad \cdot625, \quad \cdot615 \ldots\ldots\ldots \infty\}$$

3.10.6   Once again you should identify the mapping function. Do you recognise this parent set?

3.10.7   This is a TWO-TO-ONE CORRESPONDENCE because it takes *two* elements of the parent set to generate *one* corresponding element of the derived set.

3.10.8   We may meet other variations such as 'one-to-two' or 'one-to-many'; there is no practical limit.

3.11   Numeric Functions

3.11.1   In different programming languages we shall find different ways of defining or declaring functions. In our last example, for instance, we might find it coded as something like: DEF Fn d(p) = (p_i/p_{i+1}).

3.11.2   If we, as always, translate this back into plain language it says something like, "Define the function to generate any individual element, d_i, of the derived set, from the corresponding element of the parent set we divide p_i by the next element of the sequence, p_{i+1}."

3.11.3   For example, to find the fifth element of the derived set we take p_5, which is 8, and we divide it by p_6, which is 13. The fifth derived element is, therefore,

$$5/8 \text{ or } 0\cdot625.$$

3.11.4   A very common way of writing such functions in conventional mathematics is as, for example,

$$y = 2x + 7$$

3.11.5   In such traditional mathematics exercises we are often required to take a set of possible values for x and find the corresponding values for y.

3.11.6 If we set it out in the programming code form which we used in our last example we might re-state it as:

$$\text{Def fn } y(x) = 2x + 7$$

which reads, "to find any y_i we take the corresponding x_i, multiply it by 2 and then add 7."

3.11.7 When we are working such examples in the traditional way we may set the values into a table instead of in the form

$$X = \{\infty \ldots\ldots -4, 0, +4 \ldots\ldots \infty\}$$

so that we write something like

Table 3.1

IF		$x =$	-4	0	$+4$
THEN	$y = (2^*x) + 7 =$		-1	$+7$	$+15$

3.11.8 What we have, although we may never have seen it in that light before, is a set of x values mapped, in a one-to-one correspondence, onto a set of y values by the function $2x + 7$.

3.11.9 We may, in a textbook, be told that the x values are limited to integers; or are such that $-10 \le x \le +10$. This does not necessarily make it easier to select possible values for x, especially if we are testing a program routine.

3.11.10 In real life it may be the systems analyst or ourselves, as programmers, who must *set* these boundaries. We must *know* what is IN the permitted set of values for x and what is NOT. Unless we are given a specific range of values for x the set of possible values may well be infinitely large and any 'test' values which we may choose, in order to examine or display the function, are only a small sub-set. We must always remember that our sub-set is just a *representative sample of the possible values* and that the choice of test data, in particular, must always reflect that reality.

3.12 Conclusion

In a rather similar way we have seen only a sub-set of the applications of sets in this chapter.

We shall see others in practically every subsequent chapter.

$$\text{n\{applications of set theory\}} = \infty \text{ !}$$

Exercises

1) What are the two conditions governing a proper set?

2) List three collections of entities which constitute proper sets and three collections which are not. Give your reasons in each case.

3) Identify those of the following groups which are not proper sets and give your reasons:

   a) people aged over 16 years;
   b) fat people;
   c) debtors owing more that $10,000;
   d) customers living in Hong Kong;
   e) employees who work hard;
   f) Manchester United fans;
   g) a company's stock list;
   h) people who prefer Brand 'X' detergent;
   i) the test data for validating a date;
   j) real numbers.

4) Draw Venn diagrams to illustrate:

   a) $\{1,2,3,4,5\} \cup \{4,5,6,7,8\}$
   b) $\{1,2,3,4,5\} \cap \{4,5,6,7,8\}$
   c) $\{1,2,3,4,5\} \cap \{4,5,6,7\} \cap \{5,6,7,8\}$

5) A poll of 300 people boarding a Singapore flight at London Airport showed that many were continuing their journeys to other destinations. 160 were visiting Singapore; 160 were going to Malaysia; 110 were going to Hong Kong. 50 were going to Malaysia and then on to Hong Kong and 30 of these were first spending time in Singapore. 50 were merely changing flights in Singapore and going only to Hong Kong and 80 were going only to Singapore. How many were merely changing flights at Singapore to visit only Malaysia?

6) What are the mapping functions of:

   a) $\{0,1,2,3,4\} \rightarrow \{0,1,8,27,64\}$
   b) $\{33,34,35,36,37\} \rightarrow \{1,2,3,4,5\}$
   c) $\{J,F,M,A,M,J,J\} \rightarrow \{31,28,31,30,31,30,31\}$
   d) $\{0,1,2,3,4,5\} \rightarrow \{2,4,6,8,10,12\}$
   e) $\{3,5,9,13\} \rightarrow \{17,15,11,7\}$

7) Describe the sets defined by the relations:

    a) {people over 1.5m tall} $\cap$ {males}
    b) A $\cup$ B $\cup$ C
    c) {A $\cup$ B $\cup$ C}′
    d) A XOR B
    e) A NOR B
    f) $\{x: -10 \leqslant x \leqslant 10\}$

4

Number Sets

Objectives

After working through this chapter you should be able to:
- identify the different types of number and their applications.

4.1 Some Number Sets

4.1.1 We have already touched briefly on the existence of many different number sets, let us now explore this topic in a little more detail.

As I work I see that there are 6 pieces of computer equipment on my table and I may express this as:

C = {computer equipment on my work table} and n(C)= 6 or, the number of elements in C is 6.

4.1.2 To what number set does this 6 belong? Any collection of *actual entities* has a NATURAL size or quantity, whether or not we ever put it into words or symbols.

4.1.3 This has often been described, if somewhat clumsily, as the 'n-ness' of the group of entities. In strict mathematical language it may be described as the QUANTUM of the group, which is *independent of the symbols or words which we use to describe it*!

In the example shown in Figure 4.1, of horses in a field, the group has this *inherent* size, or 'number' which we happen, in English, to call 'five' but the word is only one way of describing that number.

Figure 4.1 The 'N-ness' of horses!

4.1.4 Man has invented the *word*, but the *quantum* is a *natural* attribute of the group. Not surprisingly we describe such numbers as {NATURAL NUMBERS}.

What is important to us is that *all other numbers are man-made concepts*. For example we cannot, in nature, have half a horse. In other words the central idea, the *concept*, of fractions is man-made and entirely *un*-natural.

Equally the 'five-ness' of our group of horses does not, in nature, have the LABEL 'five' much less is it identified by the numeral 5.

The words, the numerals or figures and the column system are all simply parts of a system of NOTATION by which we *describe* and *manipulate* numbers; a set of human inventions which have evolved into the commonly used system of today over many centuries.

4.1.5 From the very outset we must always distinguish between the NUMBER, the actual *quantum* or quantity, and both the SYMBOLS and the NUMBER SYSTEM which we use to *represent* that number.

4.1.6 Why do we need the words and the symbols? First and foremost in order to communicate, store and manipulate numbers representing quantities or sizes of things.

4.1.7 A secondary, but quite different application is where we also use them to identify individual positions within a sequence.

For this chapter we are concerned first with the *types* of number which we use and secondly with the systems which allow us to express and manipulate those numbers.

Since all computer systems allow us to economise on time and space by using the type of number best suited to any particular application, and since they also store and manipulate numbers according to several systems which are quite different from the one which we use in everyday life, *we* need a much clearer understanding of the nature and qualities of numbers than do ordinary users of mathematics.

Once again it is the *understanding* of principles which is important rather than traditional skills at 'doing sums'.

4.1.8 For the purpose of labelling the positions of items in a sequence, *or as a unique identifier in a set such as a file*, for instance as a firm's customer numbers, we use {ORDINAL numbers.}

In words, these are the ones we speak as 'first, second, third.....' even though we write them in numerals as '1, 2, 3.....'.

4.1.9 All other, *arithmĕtic*, numbers are within {CARDINAL numbers.}

In this case the word 'cardinal' means 'of first importance', since the first importance of numbers is to describe quantities.

4.1.10 We must be very cautious indeed here. In ordinary circumstances we can say, quite simply, "We cannot, and must not attempt to, do arithmetic with ordinal numbers". We cannot, for example, say that house number 2 + house number 3 = house number 5! We cannot even say that, because 5–2=3, there must be three houses between 2 and 5; clearly there would, at most, be two but there may be none!. Such examples quickly reveal the absurdity of the pseudo-arithmetic.

4.1.11   In our special field, however, there is no such simple separation of the ordinal numbers from the others. Within computing such numbers may well *be* subjected to a range of arithmetic manipulations, for example in the generation of check digits or in making equal subdivisions of a file by deriving true quantities, or natural numbers, from differences between customer serial numbers or dates, say.

4.1.12   The second problem for us is that within the ALU, as in storage, ordinals appear to be no different from natural numbers. They are, in fact, handled exactly as if they *were* ordinary positive integers. Once again it is we who have to understand the critical differences and ensure that, *at no time in our programming* can the ordinals be wrongly included in calculations. The computer doesn't know the difference; therefore *we must*!

4.2   Subsets of Cardinal Numbers

4.2.1   So far, as subsets of {cardinal numbers}, we have identified {natural numbers} and we have, in Chapter 3, touched upon {integers} and {reals}.

How large is the set of natural numbers? What can it *not* contain? (Clearly, although it has the same elements, it cannot contain the {ordinal numbers}!)

These are the questions which we must always ask.

How many stars in the sky? Stars are certainly natural objects and have a natural 'n-ness'. Astronomers, for their special study, do make estimates of n{stars} – but if we started counting them we should have a quite hopeless task; there are simply too many and whatever number we reached in our counting someone else could always say "..and one more!"

4.2.2   For ordinary mortals that is probably as good a working model of infinity as any. Theoretically we *could* count the members of any natural set but in many cases, such as the stars, we should always be able to say, "..and one more." {Natural numbers} must then be infinitely large and if we were listing it we could write:

$$\{1, 2, 3, 4, 5, \ldots\ldots\infty\ \}.$$

4.2.3   This set coincides with {+ve integers} since all its elements are whole numbers only, without fractions, and they are all >0.

4.2.4   0 is not, of course, a number at all; it is really the absence of number. It is useful, nonetheless to treat it as the ORIGIN of all counting and of the number line.

4.2.5   Nature, however, has no 'minus' quantities, we cannot naturally have fewer than *no horses*. It is man who has invented the ideas of 'below sea-level; below zero; having an overdraft; owing a debt to someone; a negative balance on an account' and all the many conditions which we describe by negative (–ve) numbers.

4.2.6   So the set of *all* integers, as opposed to {+ve integers} has {natural numbers} as one of its subsets. Thus:

{natural numbers} ≡ {+ve integers} but
{natural numbers} ⊂ {integers}.

The two number lines are also quite different, as can be seen in Figure 4.2:

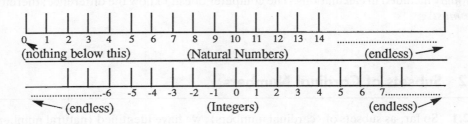

Figure 4.2   The two sets as number lines

and reflect the great extension of {Integers} into the *un-natural*, negative integers.

4.3   Fractional Numbers

4.3.1   If we need to deal with the *real* world, however, we may well need to measure, compare or account for parts of units. We could not, for example, measure distances only in whole metres – think of what our problems would be if we could only buy shoes which were 1m long!

4.3.2   Unless we led very leisurely lives it would also be extremely inconvenient if all we knew of the departure time of a train was that it is, "..sometime between three and four o'clock in the afternoon"; or if, even for a family of two people, we could only buy tea in complete kilos.

4.3.3   In the real world we simply have to sub-divide units and we express all *measurements* in quantities which reflect this fact. This means that, although we can *count* anything in natural numbers and express shortages of *items* in negative integers we must have a third set of numbers in which we can express, *not only* whole quantities *but also* any parts of units, no matter how small.

4.4   Real Numbers

4.4.1   This need reflects the demands of the *real* world and we call the resulting numbers {REAL numbers}.

4.4.2   Let us take a magnifying glass to the number line as shown in Figure 4.3:

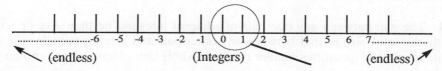

(endless)   (Integers)   (endless)

Figure 4.3   'Closing in' on one part of the number line

and look at what, in the *real* number world, comes into the gap between *any pair* of integers.

In Figure 4.4 we see three 'stages' of magnification of the 'gap' between 0 and 1 on the INTEGER scale which holds many, indeed an infinite number of, other values on the REAL NUMBER scale. In each case the thick line above the scale denotes the quantity which we are seeking to identify. At stage *a*) we see the greatest degree of precision which we can arrive at using integers.

Stage *b*) shows us that we can considerably increase the precision if we sub-divide our UNIT into ten equal parts or fractions, in this case, tenths.

By sub-dividing each tenth into tenths we can reach still further precision and identify the quantity as >24/100 and <25/100. We can, at will, go on sub-dividing the 'gap' between any pair of integers into any level of detail that we choose.

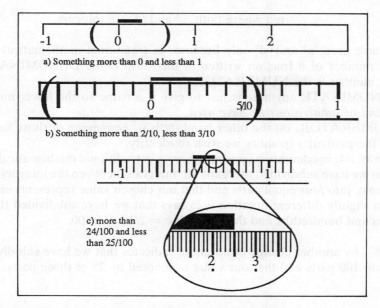

a) Something more than 0 and less than 1

b) Something more than 2/10, less than 3/10

c) more than 24/100 and less than 25/100

Figure 4.4   Filling in, or opening out, the intervals between integers

4.4.3   In this instance I happen to have chosen to sub-divide into tenths, hundredths and so on. I might just as readily have chosen to sub-divide the unit into halves, quarters, eighths, sixteenths...... or any other kind of part which suited me. The key word is 'suited'; we should choose whichever sub-divisions are appropriate to the task in hand and be able to change from one scale to another at need.

The *values* are all there, only the *names* of the subdivisions vary. We can as readily write or use, as it suits us, ¼, 0·25, 25/100 or 25%; they all mean exactly the same position in the 'gap' between 0 and 1 in our {Real Numbers}.

4.4.4   The gap between any one pair of adjacent integers, be they +ve or −ve, is exactly the same size as that between any other pair and actually contains the same infinite number of quantities smaller than one whole unit, whatever our chosen scale may be.

4.4.5   The type of fractions by which we *describe* those interim values is only a matter of name. All the actual values concerned are the same RATIONAL NUMBERS.

4.5   Rational Numbers

4.5.1   This is actually another of those rather strange labels whose roots lie in the days when only a very few specialists actually studied mathematics and, like all specialists, they spoke their own jargon. What it means is that the value can be expressed as a 'fraction' of the form

n/d where both 'n' and 'd' are integers.

4.5.2   I have used 'n' and 'd' only because, in traditional mathematical language, the lower number of a fraction written in this form is the DENOMINATOR and the upper number is the NUMERATOR.

To DENOMINATE anything is just to give it a name so the lower number tells us *what kind of subdivisions we have used*.

The NUMERATOR, on the other hand, tells us *how many* of those subdivisions represent the particular quantity we wish to identify.

In this way 1/4, in computing convention, or $\frac{1}{4}$ in traditional mathematical notation, tells us that we have subdivided the 'gaps' or 'intervals' between the integers in the real number scale, into *four* equal parts and that our chosen value represents *one* of those parts. In a slightly different notation 0·25 says that we have subdivided the interval into tenths and hundredths and that the value = 2/10 + 5/100.

4.5.3   25%, by another change of notation, indicates that we have sub-divided each interval into 100 parts and that our value is denoted by 25 of those parts.

4.5.4 Every 'finite' value less than one whole, ie a value whose size or 'magnitude' we can represent in such ways is

∈ {Rational Numbers}

4.5.5 All rationals may be represented in many different ways depending upon the chosen denominator and not only does

{ ¼   ½   ¾ }
↓   ↓   ↓
{ 0.25 0.5 0.75 }
↓   ↓   ↓
{ 25% 50% 75% }

but   { ¼   ½   ¾ }
↓   ↓   ↓
{ 3/12 6/12 9/12 }

or any similar mapping with any other denominator.

4.5.6 Remember as you practise such mappings that both numerator and denominator must be integers.

4.6 Irrationals

4.6.1 There are values, which we work with quite commonly, which are outside {rationals} and hence are within {irrational numbers}

In this case the word 'irrational' does not imply 'silly' or 'eccentric', as it often does in everyday English. All it says is NOT (in the set of) RATIONALS.

4.6.2 Examples are all around us and again we must be wary of them in computing since, within a program, they can, if we are not careful, create an endless loop. Not infrequently when a machine has appeared to 'freeze' we have given it an impossible task such as calculating an irrational number!

4.6.3 A very common irrational is the value of π. No matter how long we keep on calculating we never seem to come to an end where n/d can be written in integers, not even at half a million places of decimals have we found a *finite* value for π ! Almost equally common are values like √2. This too, no matter how diligently calculated, has never yet come to a finite n/d. For π we may, for elementary calculations, use 22/7, or, for greater precision, 3·141915 but neither is a finite, rational, value even if they have that appearance.

4.6.4 {Irrationals} has a very special significance for us since any given value which we may have to calculate, eg, $\sqrt{2}$; $\sqrt{3}$; $\sqrt{5}$, will not be *identified* as an irrational and we may be unpleasantly surprised by runtime problems with a program.

4.6.5 The second reason is that whenever we call for such a value in a program, if our translator program has something equivalent to 'sqrt(2)', for example, the computer does not 'look up' a table of such values but must calculate them afresh by an internal sub-routine. Once again, we need to be aware of the *limitations* of our systems.

4.7 Levels of Precision

4.7.1 Some interpreters or compilers confront us with two more number classes, {single-precision reals} and {double-precision reals}. These two, purely computer-related sets, highlight an inherent characteristic of the computer. We may 'juggle' its processor to give increased precision *but* if we want increased precision in calculations we must sacrifice speed.

4.7.2 Quite commonly real numbers will be described in the particular computer manual as 'calculated to 8 and published to 7 significant figures.' Thus a value for π would be calculated as 3·1415926 and shown on the screen, or other output, as 3·141593.

4.7.3 If we need a greater degree of refinement of their value than this we may, by invoking 'double-precision', cause the computer to artificially extend the calculating capacity of the ALU and calculate to 16, publishing to 15, significant figures.

4.7.4 In this case the actual process of computation is considerably slower but the corresponding values for π would be 3·141592653589793, output as 3·14159265358979.

4.7.5 Once again the important thing is for us to realise the implications of very precise calculation of numbers and the cost, in processing time, of doing so. In speed terms {integers} allows the fastest time, {single-precision reals} the next fastest with {double-precision reals} coming a poor third.

4.8 Square Numbers

4.8.1 Here are some more mappings; derive the mapping function and name set B in each case.

i)   A = { 1, 2, 3, 4, 5, ... ∞}
   ↓   ↓   ↓   ↓   ↓
   B = { 1, 4, 9, 16, 25, ... ∞}

ii)   A = { 1, 2, 3, 4, 5, ... ∞}
   ↓   ↓   ↓   ↓   ↓
   B = { 1, 8, 27, 64, 125, ... ∞}

Can you see how example i) relates to:

iii)   A = { 1·5,   2·5,   3·5,   4·5, ... ∞}
   ↓   ↓   ↓   ↓   ↓
   B = { 2·25, 6·25, 12·25, 20·25, ... ∞}?

4.8.2 Set i) A is {natural numbers} and
Set i) B is {natural square numbers},

which is, it appears, another of the 'silly' descriptions since the whole idea of a 'square' is man-made and there are precious few squares, if any, to be found in nature!

4.8.3 The explanation of this apparent foolishness is that some natural groupings can be put into the form of squares which have sides whose lengths are natural numbers. It is easiest to see if we use circles, or counters, to represent the natural entities.

In Figure 4.5 the lengths of side of the successive squares is indicated and the growth patterns should be clear. If our natural square number is 25, ie can be represented as a square containing 25 counters, then its ROOT is the length of its side, 5.

By the inverse process a square with sides which are 5 units long will contain 25 counters. In conventional mathematical symbolism, 5^2, implying 5 multiplied by itself, or 5*5, = 25. (In computer notation, 5↑2 or 5**2.)

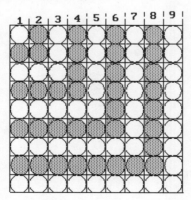

Figure 4.5 Natural squares

4.9 Conclusion

There are many more number sets, but for our purposes everything we are likely to meet in everyday computing will probably be in one of those which we have discussed here or in one of their subsets. You should, however, get as much practice as you are able in relating numbers, according to the application, to their relevant set or sets. Such practice should include declaring the variables, defining acceptable ranges and setting up 'GO and NO-GO' test data suitable for use in a real program.

Exercises

1) Which of the following are ∉ {natural numbers}:
   4; 4.2; −5; 7; 14.9; √2; √4; 0.

2) List six relationships between {natural numbers}, {integers} and
   {real numbers}.

3) Complete the mappings:

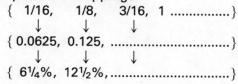

     { 1/16,   1/8,   3/16, 1 ....}
       ↓   ↓   ↓
     { 0.0625, 0.125, ...}
      ↓   ↓   ↓
     { 6¼%, 12½%, ...}

4) List the members of {√1; √2; √3; √4; √5; √6} ∩ {irrationals}.

5) List for x ≤ 730, {∛x} ∩ {natural numbers}.

5
A First Look at Number Systems

Objectives

After working through this chapter you should be able to:
- understand the column system and the number base;
- see the relevance of base 2 and base 16;
- work with modular arithmetic.

5.1 Some Number Systems

5.1.1 We have four tools for *describing* numbers of all sizes:

a fundamental counting group or NUMBER BASE;

a set of symbols, $\{0, +, -, \cdot, \infty\}$;

a set of numerals, in our case, $\{1, 2, 3, 4, 5, 6, 7, 8, 9\}$, which we have already looked at;

and a COLUMN SYSTEM which allows us to combine numerals and zeros to denote any conceivable number no matter how large or how small.

5.1.2 These things taken together must be counted among man's most powerful inventions. The counting group, and hence the set of numerals we use, is determined by the fact that the human race happens to have evolved with four fingers and one thumb on each hand. When early man (and very young children even today), counted objects he did so by mapping the objects onto his fingers, or putting them into one-to-one correspondence.

The maximum counting group in this case is when all fingers and thumbs have been matched with objects. We then need either another pair of hands or some way of 'noting down' that we have accounted for one 'double handful' of fingers.

5.1.3 The Roman notation reflected this mapping exactly, even putting two 'V's together to make 'X' the symbol for a double handful, as shown in Figure 5.1. It is from such elementary attempts to record and communicate numbers, all over the world and in many different cultures, that the universal number system, in BASE 10, has evolved.

We indicate numbers written in this system as, for example, 365_{10}, by adding the *subscript* 10 at the end of the number. The Roman system, however, lacked any

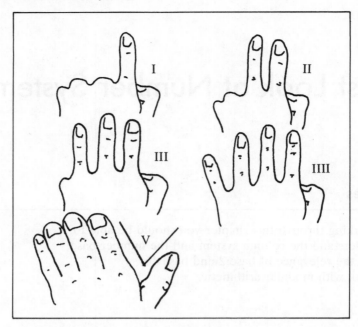

Figure 5.1   The 'finger' mapping of the Roman notation

indicator for zero and a column system was not developed. As a result the notation was cumbersome and inflexible, even though simpler and easier to write than the earlier Egyptian or Greek notations.

5.1.4   Our current numerals are derived from Hindu–Arabic origins and, with the essential contribution of the symbol '0' to represent 'no number' we eventually developed our modern notation. In this the column in which a numeral is written determines its *magnitude relative to the number of units it indicates*. The numeral 5 indicates, for example a group of the size of our earlier group of horses. If the same numeral is written in the next column to the left the number which it indicates is ten times as large – or the equivalent of gathering together into one group *ten* groups of horses of the size of that in Figure 4.1. Put together, the numerals, zero and the columns give us a truly powerful and flexible system.

5.1.5   The concept of position, or column, denoting magnification of a value is closely akin to that of a variety of lenses giving us a variety of 'powers of magnification'. A '5 power', or 5×, lens enlarges objects to 5 times the size at which they appear to the naked eye; a 10× or 10 power lens will make things appear 10 times as large. We express the 'power' of a position in the columns to 'enlarge' a number as 10* or 10*10*, and so on. As always in mathematics there is a shorthand for such notations. 10*10*10*10*10*10*10*10*, as in the leftmost column of Table 5.1, is abbreviated to 10^9 which we say in words as "10 *to the power nine*" or, "10 *to the ninth (power)*."

5.1.6 In this table (which has no end at the left side where the columns can go on for ever and in which the numbers are always ten times as large as in the preceding column) we see various notations, both in symbols and in words:

Table 5.1

	10^9	10^8	10^7	10^6	10^5	10^4	10^3	10^2	10^1	10^0
i)										
ii)	10*10* 10*10* 10*10* 10*10* 10*	10*10* 10*10* 10*10* 10*10*	10*10* 10*10* 10*10* 10*	10*10* 10*10* 10*10*	10*10* 10*10* 10*	10*10* 10*10*	10*10* 10*	10*10*	10*	1*
iii)		MILLIONS			THOUSANDS			UNITS		
iv)	U	H	T	U	H	T	U	H	T	U
a)			3	0	5	7	2	3	2	7
b)					4	2	3	5	2	4

5.1.7 Taking the various levels of the table in turn:

i) This is the POWERS OF TEN notation with NO magnification in the rightmost table, the UNITS column, but with each shift along the columns to the left adding another 'ten times' to the degree of 'enlargement' of the number.

ii) In this section the number of multiplications by ten is detailed. Thus in the 10^6 column, for example, any numeral, let's say 3, represents $3*10*10*10*10*10*10 = 3*10^6$ or three million; 3,000,000.

iii) In the next 'layer' we see the major words for describing large quantities. These are the English sub-divisions of numbers and these do vary from country to country.

iv) In this repetition of H, T, U we see one of the verbal descriptions which most children learn quite early in their lives. Hundreds, Tens and Units, or, more sensibly, Units, Tens and Hundreds. In English verbal descriptions this pattern is repeated so that when combined with level *iii)* descriptions we continue into Thousands, Tens of Thousands, Hundreds of Thousands, Millions, Tens of Millions, Hundreds of Millions and so on.

5.1.8 At this stage you should look at rows a) and b) and:

i) both say, in words, and write down in figures, the overall value of each of the two numbers;

ii) write down the value of *each numeral* treating the UNITS column as column 1; for example in a) the 3 in column 3 is worth 300. What is the value of the 3 in b)?

iii) compare the sizes of like numerals in different columns by their powers; for example the leftmost 3 in a) = $3*10^9$ and the rightmost 3 is $3*10^2$, thus the leftmost is 10^7 times as large as the rightmost. Make *quite sure* that you understand the reasoning here.

iv) compare the sizes of *unlike* numerals in the same way. For example: the rightmost 2 in b) is $2*10^1$ and the leftmost is $4*10^5$; the latter is $2*10^4$ times as large as the former.

5.1.9 Again, it is important to discuss the principles and to practise with plenty of examples to make sure that you are confident with these notations. Showing a small *superscript* numeral to denote the power of a number is known as INDEX NOTATION and the examples in *iii*) and *iv*), combining the number of UNITS with the INDEX is widely known as SCIENTIFIC NOTATION. All such notations are of great importance to us in computing where they are commonly written as, for example, $10 \uparrow 9$, since translator programs rarely recognise sub- or super-scripts.

5.2 Other Bases

5.2.1 This system should, strictly speaking, be described as the DENARY system but in common language it is usually referred to as the DECIMAL system although the decimal notation did not evolve until many generations later. The biological accident of our having ten fingers and thumbs, digits, might just as easily have seen us evolve with eight on each hand!

5.2.2 We should then, equally naturally, have developed numbers in BASE SIXTEEN. Until very recently such thoughts merely gave mathematians some harmless fun! All that has changed with the introduction of two-state, digital computers (as distinct from 'analogue' computers) in which base 16 numbers are widely used.

5.2.3 Base two numbers need only the numerals 1 and 0 and its column system, essentially exactly the same in nature, sees enlargement by powers of two instead of powers of ten. This is shown in Table 5.2.

Table 5.2

	2^9	2^8	2^7	2^6	2^5	2^4	2^3	2^2	2^1	2^0
i)										
ii)	2*2*2*2*2* 2*2*2*2*	2*2*2*2*2* 2*2*2*	2*2*2*2*2* 2*2*	2*2*2*2*2* 2*	2*2*2*2*2	*2*2*2*2*	2*2*2*	2*2*	2*	1*
iii)	1024's	512's	256's	128's	64's	32's	16's	8's	2's	U
a)			1	0	0	1	1	1	0	1
b)					1	1	1	0	0	1

5.2.4 Try to do, with these numbers, the same exercises as before, in section 5.1.8. You may conclude that having words like 'millions' offers some advantages!

5.2.5 We shall need to be fairly fluent in working with the BINARY numbers in practice and need experience to develop confidence.

5.3 Hexadecimal Numbers

5.3.1 Describing values in binary numbers, although essential at the very heart of the computer, does take up a great deal of space, especially when we write them or try to manipulate them on paper. For these and many other reasons, but most of all because it is easy to convert to and from base two, we do actually make a great deal of use of numbers in base 16, or HEXADECIMAL numbers, commonly shortened to HEX.

5.3.2 Unfortunately we have only nine numerals and zero in our base ten system whilst we need no fewer than *fifteen* numerals, to accompany zero, in HEX. It has become accepted practice, since they are already available both on the keyboard and in computer code systems, to use A_{16} for 10_{10}, B_{16} for 11_{10}, C_{16} for 12_{10}, D_{16} for 13_{10}, E_{16} for 14_{10} and F_{16} for 15_{10} and to 'advise' the computer that these are not part of an alphanumeric string, by adding &H before the 'numerals' (since A to F are now also numerals) as in &H1AF, or an 'H' after the number as in 1AFH. Unfortunately, the notation is not standard in computer literature and we must be prepared to work with whatever form is used in the compiler, interpreter or operating system with which we are concerned.

We do not use these labels, however, in ordinary arithmetic with base 16 numbers but it *is* wise to get into the habit of using the base-indicating subscripts. On the face of it there is nothing to tell us, if we *are* using different bases, what base a number such as 479 may be in although 4F7 *might* be assumed to be in base 16 but we should never rely on such supposition.

In computing, as in mathematics, it is sound practice to assume that if anyone *can* misunderstand or misinterpret what we have written it is inevitable that, sooner or later, someone *will*.

5.3.3 The column system gives us a pattern, shown in Table 5.3:

Table 5.3

i)	16^9	16^8	16^7	16^6	16^5	16^4	16^3	16^2	16^1	16^0
ii)	16*16* 16*16* 16*16* 16*16* 16*	16*16* 16*16* 16*16* 16*16*	16*16* 16*16* 16*16* 16*	16*16* 16*16* 16*16*	16*16* 16*16* 16*	16*16* 16*16*	16*16* 16*	16*16*	16*	1*
iii)							4096's	256's	16's	1's
a)			3	A	5	F	2	F	A	7
b)					4	B	E	E	F	8

5.3.4 Once again you should try all the exercises in 5.1.8 to make sure that you are comfortable with using numbers written in base 16.

5.4   Converting from Base to Base

5.4.1 Today, many computer translators and even many pocket calculators are programmed to convert numbers from one commonly used base to another. It is useful, however, to be able to do so by pencil-and-paper methods and a good start is to make sure that you can list our {natural numbers} in at least our three 'standard' notations.

5.4.2 It is also very useful to look at the addition and multiplication squares for different bases. These, if you have not met them before, are an excellent way of examining the behaviour of numbers and we will look at examples later in the chapter.

5.4.3 To take a listing of the counting, or natural, numbers first. The Binary numbers are in '8 bit' form.

DENARY	BINARY	HEXADECIMAL	OCTAL (Base 8)
1	00000001	1	1
2	00000010	2	2
3	00000011	3	3
4	00000100	4	4
5	00000101	5	5
6	00000110	6	6
7	00000111	7	7
8	00001000	8	10
9	00001001	9	11
10	00001010	A	12
11	00001011	B	13
12	00001100	C	14
13	00001101	D	15
14	00001110	E	16
15	00001111	F	17
16	00010000	10	20
17	00010001	11	21
18	00010010	12	22
19	00010011	13	23
20	00010100	14	24
21	00010101	15	25
22	00010110	16	26

23	00010111	17	27
24	00011000	18	30
25	00011001	19	31
26	00011010	1A	32
27	00011011	1B	33
28	00011100	1C	34
29	00011101	1D	35
30	00011110	1E	36
31	00011111	1F	37
32	00100000	20	40

5.4.4 I have, to give you something a little new to think about, included some numbers in base 8, which has some applications in computing. Try to develop the column system, index notation and so on for this base as well as doing the exercises in 5.1.8.

5.4.5 When we want to change a base 10 number to, say, base 8, we do a repeated division by 8 and preserve all 'remainders'.

5.4.6 If we wish to convert a base 'n' number back to base 10 we do the converse, digit-by-digit multiplication by 'n' and include all partial products, ie all intermediate results.

5.4.7 Convert 475_{10} to its base 8 equivalent:

$$
\begin{array}{r|l}
8 & 475 \\
8 & 59 \text{ r } 3 \\
& 7 \text{ r } 3
\end{array}
$$

and we read from the last integer, 7, *right and upwards* to include all the remainders: $475_{10} = 733_8$

5.4.8 To convert 733_8 to base 10:

$$
\begin{array}{ccc}
7 & 3 & 3 \\
*8 \rightarrow & +56 \rightarrow & +472 \\
56 \rightarrow \uparrow & \dfrac{59}{} \uparrow & \overline{475} \\
& *8 \uparrow & \\
& \overline{472} \rightarrow &
\end{array}
$$

and we can confirm that $733_8 = 475_{10}$.

5.4.9 The same process for Binary conversions is just a little more laborious and we need to take extra care not to lose any remainders – especially the zeros!

$$
\begin{array}{r|l}
2 & \underline{475} \\
2 & \underline{237}\,\text{r}\,1 \\
2 & \underline{118}\,\text{r}\,1 \\
2 & \underline{59}\,\text{r}\,0 \\
2 & \underline{29}\,\text{r}\,1 \\
2 & \underline{14}\,\text{r}\,1 \\
2 & \underline{7}\,\text{r}\,0 \\
2 & \underline{3}\,\text{r}\,1 \\
  & 1\,\text{r}\,1
\end{array}
$$

If we read from the bottom upwards this value is 111011011 and, by referring back to our BINARY column system we can evaluate this, from the units column, as $(1 + 2 + 8 + 16 + 64 + 128 + 256)_{10} = 475_{10}$

5.4.10 Once again, if we reverse the process and do the continuous 'multiply by two' routine we see:

1	1	1	0	1	1	0	1	1
* 2	2	6	14	28	58	118	236	474
2	3	7	14	29	59	118	237	475
	* 2	* 2	* 2	* 2	* 2	* 2	* 2	
	6	14	28	58	118	236	474	

5.4.11 There are several versions of these processes but these are perhaps the simplest. As ever, they need practice to develop confidence.

5.4.12 Conversions from binary to hex. and vice versa are somewhat simpler since one BYTE, of eight BITS (Binary DigITS) consists of two NIBBLES of four BITS each and one hex digit can be represented by four bits. If we look again at our counting numbers, with the Binary numbers in '8 bit' form and separate these bytes into nibbles:

DENARY	BINARY	HEXADECIMAL
1	0000 0001	1
2	0000 0010	2
3	0000 0011	3
4	0000 0100	4
5	0000 0101	5
6	0000 0110	6
7	0000 0111	7
8	0000 1000	8
9	0000 1001	9
10	0000 1010	A
11	0000 1011	B

12	0000 1100	C
13	0000 1101	D
14	0000 1110	E
15	0000 1111	F

We can see that only the right-hand nibble is used for numbers up to 15_{10} or F_{16}, the left-hand nibble, being entirely made up of zeros, is merely space-filling, is worth nothing and may be ignored at this stage.

This means that, if we separate out any binary integer into nibbles from the units column leftwards, we can transpose each nibble directly into its hexadecimal equivalent.

5.4.13   For 111011011_2 then, we may write this (in nibble form and using 'filler' leading zeros to complete the leftmost nibble) as:

$$0001 \quad 1101 \quad 1011, \text{ which we can express as:}$$

$$1 \quad D \quad B, \quad \text{respectively, in base 16, or } 1DB_{16}.$$

This would give us $1*256 + 13*16 + 11 = 475_{10}$

5.4.14   We can, of course, do the repeated division routine or the repeated multiplications, for base 16 as well as any other. Try it. This may be a good argument for learning our sixteen times table!

5.5   A System for all Real Numbers

5.5.1   In the early seventeenth century, the next revolutionary change to the number system took place. If we look again at the model in Table 5.4 we may, instead of seeing each leftwards move in the columns as another 'multiply by ten', *start* from the left and perceive that we perform the *inverse* operation in the *inverse* direction.

5.5.2   For each move to the *right* in the columns the value of a numeral is 'diminished' by ten times or is 1/10 the size of its left-hand neighbour. This is shown in Table 5.4 in which it is possible to identify any value smaller than 1 by a mirror image of the integers section of the system. What is the number at a) now?

5.5.3   As with so many such advances a new symbol had to be devised because we have always used the units column as the reference for the ultimate size of a number and whereas this is clear whilst we keep the full column structure and its labelling, it disappears when we 'rub out' those guides and merely write the number down without them. What became necessary, immediately this extension to the column system was conceived, was a foolproof indicator of *the position of the units column* in order to make the value of the number in b) in Table 5.1 clearly different from the number at b) in Table 5.4.

Table 5.4

i)	10^4	10^3	10^2	10^1	10^0	10^{-1}	10^{-2}	10^{-3}	10^{-4}	10^{-5}
ii)	10*10* 10*10*	10*10* 10*	10*10*	10*	1*	÷10	÷10÷10	÷10÷10 ÷10		
		MILLIONS			THOUSANDS			UNITS		
iv)	U	H	T	U	H	T	U	H	T	U
a)			3	0	5	7	2	3	2	7
b)					4	2	3	5	2	4

The solution which emerged was to use a dot to the right of the relevant numeral, *to designate the units column* for any number.

If we write the two numbers using this notation we have, from Table 5.4, 30572327 whereas here we have 305·72327, a very different quantity.

5.5.4   By an extension of the 'each column to the right makes the value one tenth as large' we can denote the new values by *subtracting 1 from the index for each column that we move to the left*. Since $0 - 1 = -1$ the index notation for the tenths column is 10^{-1}, the hundredths column becomes 10^{-2} and so on.

5.5.5   The new system was described as the DECIMAL system, meaning tenths-based fractions, as opposed to the DENARY system meaning 'tens based'. As we saw earlier this distinction has become blurred and most computer manuals refer to decimal, rather than denary, values.

What it gave us was an immensely powerful up-grade of our number system so that, as the columns extend infinitely far in *each direction*, we are now able to write *any real number*, no matter how large or how small, in the same notation rather than using two notations as in, for example, $2\frac{3}{8}$.

5.5.6   We can and do apply the same principle to the binary system to create a BICIMAL (fraction) system. In this case the columns become $\frac{1}{2}$'s, $\frac{1}{4}$'s, $\frac{1}{8}$'s and so on.

We could do likewise with the hexadecimal system but we normally have no need to do so and it is certainly beyond our present scope.

When, in a later chapter, we look at the representation and manipulation of numbers in the computer we shall rely quite heavily on this material so it is well worthwhile trying to be confident in our understanding and execution of these ideas.

5.6   A Very Different Number System

5.6.1   The use of an OPERATION or TABLE SQUARE is one way to look at the patterns of number behaviour which arise from the combination of the number system and the laws. Let us, for simplicity's sake, only consider the 'counting up' operations of + and * and the values 1 to 10 only, together with 0, so that we can also see the effects of the laws relating to zero.

To clarify two very important but quite basic ideas by, once again, translating our symbols back into words: 3 + 4, (or any other a + b), written in full says something like, "Starting at zero count up to 3 and then count up four more. What is the position reached on the number line?" You should make sure that you can illustrate this, as well as other examples, with a sketch of the process.

3*4, (or any a*b), on the other hand, says, "Start at zero and count up three lots of four. Where is the final position on the number line?" Once again you should be sure that you can illustrate such operations.

NOTE: wherever we see the + sign we can replace it with the word 'and', with the word 'of' being a similar replacement for the * symbol.

Both the symbols are OPERATORS, sometimes shown as ⊕ and ⊗ respectively, indicating just what we must do with the two numbers which they indicate.

5.6.2 In this operations square (Table 5.5), we see the complete 'tabulation' of all possible results of any two numbers in base ten, up to 10, being added.

Look for patterns and any other features which seem important, discuss them with your colleagues and try to link them with the laws which we discussed in Chapter 1.

Table 5.5

⊕	0	1	2	3	4	5	6	7	8	9	10
0	0	1	2	3	4	5	6	7	8	9	10
1	1	2	3	4	5	6	7	8	9	10	11
2	2	3	4	5	6	7	8	9	10	11	12
3	3	4	5	6	7	8	9	10	11	12	13
4	4	5	6	7	8	9	10	11	12	13	14
5	5	6	7	8	9	10	11	12	13	14	15
6	6	7	8	9	10	11	12	13	14	15	16
7	7	8	9	10	11	12	13	14	15	16	17
8	8	9	10	11	12	13	14	15	16	17	18
9	9	10	11	12	13	14	15	16	17	18	19
10	10	11	12	13	14	15	16	17	18	19	20

5.6.3 Table 5.6 also shows operations on numbers in base 10 and in quite ordinary, everyday circumstances, but with important differences. Can you offer any explanation of the changes?

Table 5.6

⊕	0	1	2	3	4	5	6	7	8	9	10
0	0	1	2	3	4	5	6	7	8	9	10
1	1	2	3	4	5	6	7	8	9	10	11
2	2	3	4	5	6	7	8	9	10	11	0
3	3	4	5	6	7	8	9	10	11	0	1
4	4	5	6	7	8	9	10	11	0	1	2
5	5	6	7	8	9	10	11	0	1	2	3
6	6	7	8	9	10	11	0	1	2	3	4
7	7	8	9	10	11	0	1	2	3	4	5
8	8	9	10	11	0	1	2	3	4	5	6
9	9	10	11	0	1	2	3	4	5	6	7
10	10	11	0	1	2	3	4	5	6	7	8

5.6.4　Now we have Table 5.7 showing the first few multiplication operations in base 10. In all the cases we have looked at the numbers could, of course, go on to infinity in both fractions and negative numbers so we are looking only at a SAMPLE SPACE of the whole of our number system.

What is helpful is that the patterns and the effects of the laws are the same over the whole number system so it is worth examining these tables very carefully to see what they tell us.

Table 5.7 is clearly different from Tables 5.5 and 5.6. What are the reasons for those differences? Are there any similarities; what are they and what do they indicate?

Table 5.7

⊛	0	1	2	3	4	5	6	7	8	9	10
0	0	0	0	0	0	0	0	0	0	0	0
1	0	1	2	3	4	5	6	7	8	9	10
2	0	2	4	6	8	10	12	14	16	18	20
3	0	3	6	9	12	15	18	21	24	27	30
4	0	4	8	12	16	20	24	28	32	36	40
5	0	5	10	15	20	25	30	35	40	45	50
6	0	6	12	18	24	30	36	42	48	54	60
7	0	7	14	21	28	35	42	49	56	63	70
8	0	8	16	24	32	40	48	56	64	72	80
9	0	9	18	27	36	45	54	63	72	81	90
10	0	10	20	30	40	50	60	70	80	90	100

5.6.5 Now let us look at a multiplication square, also in base 10, but which again differs from Table 5.7. Can you offer any suggestions as to why?

Table 5.8, which follows, shows the same first few multiplications in a base ten number system, of everyday familiarity, also behaving in ways totally unlike the numbers in Tables 5.5 and 5.7.

Table 5.8

(*)	0	1	2	3	4	5	6	7	8	9	10
0	0	0	0	0	0	0	0	0	0	0	0
1	0	1	2	3	4	5	6	7	8	9	10
2	0	2	4	6	8	10	0	2	4	6	8
3	0	3	6	9	0	3	6	9	0	3	6
4	0	4	8	0	4	8	0	4	8	0	4
5	0	5	10	3	8	1	6	11	4	9	2
6	0	6	0	6	0	6	0	6	0	6	0
7	0	7	2	9	4	11	6	1	8	3	10
8	0	8	4	0	8	4	0	8	4	0	8
9	0	9	6	3	0	9	6	3	0	9	6
10	0	10	8	6	4	2	0	10	8	6	4

5.6.6 Did you get any ideas about the numbers in Table 5.6? Do your conclusions from that investigation hold good for Table 5.8?

In case you are still wondering let us write, in words, a couple of very ordinary questions.

a) If I start work at seven o'clock in the morning and work for nine hours at what time will I finish work?

b) If school starts at nine o'clock in the morning and we take seven three-quarter hour lessons with a three-quarter hour lunch break when will school end?

5.6.7 We can model a) as $7 + 9 = 4$ and
b) as $9 + (8*\frac{3}{4}) = 3$
which, compared with our ordinary arithmetic, seems very odd.

5.6.8 The explanation for the oddness of our results, and for those in Tables 5.6 and 5.8 is that the number system of the clock face ends at 12 but also begins again, at the same point, at 0!

This means not only that $12 = 0$ but that, although we can go on 'going round' the number system for ever, each time we count up to 12 it actually equals 0 and $n*0 = 0$.

This, in turn, means that the maximum possible quantity which we can express in this system is $11 \cdot 9$ recurring. Our ordinary, base 10 system goes on for ever, or is an INFINITE system, because the columns may be extended indefinitely in either direction or into the −ve numbers (see Figure 5.2).

5.6.9 The numbers in our clockface **arithmetic**, however, must be less than or equal to 12. This is a FINITE SET **and**, not surprisingly we refer to FINITE ARITHMETICS. One consequence is that $2475_{10} = 3_{\text{mod}12}$. If, for any purpose, we need to account for all the twelves which we have 'thrown away' we must do so by another mechanism.

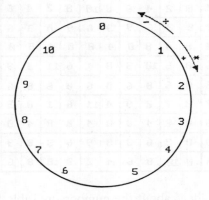

Figure 5.2   Clockface arithmetic

5.7   Finite Arithmetics

5.7.1 The size of such a group is also described as its MODULUS, (or MOD for short), and hence we talk of MODULAR ARITHMETICS. We may indicate MODULAR numbers by a subscript, for instance $3_{\text{mod}12}$ is quite a different number from 3_{10} since it can be the result of an infinite number of different operations on integers.

From the tables above what subsets of {integers} can be combined under addition or multiplication to produce 3_{10}? The normal clock face actually shows two modular systems. Can you identify them?

Table 5.8 shows that there are a great many repetitions in the mod 12 operations. What happens if we use a PRIME number MODULUS?

5.7.2 An 11 hour clock-face is shown in Figure 5.3 and part of its associated multiplication square is in Table 5.9.

Complete the square and compare it carefully with the results in Table 5.9.

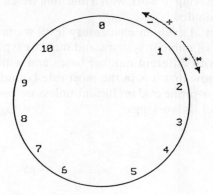

Figure 5.3 A prime number modulus

Table 5.9

(×)	0	1	2	3	4	5	6	7	8	9	10
0	0	0	0	0	0	0	0	0	0	0	0
1	0	1	2	3	4	5	6	7	8	9	10
2	0	2	4	6	8	10	1	3	5	7	9
3	0	3	6	9	1	4	7	10	2	5	8
4	0	4	8	1	5						
5											
6								9			
7				10							
8											
9											
10				8							

5.7.3   Continue the table to column and row 10 and study the results.

Discuss your conclusions fully and make sure, by trying other MODs, that you have fully grasped the principles of FINITE ARITHMETICS; they do have many uses in computing.

You should also practise converting numbers from their base ten form to a variety of MODs.

5.8 Conclusion

Whilst it is tempting to leave number theories to the mathematicians, we are in the very strange position of having to work with a machine which is both very sophisticated and alarmingly simple-minded.

In order to program it at even an elementary level we must have a good working grasp of the very basics of number systems and number types. We must also be able to function effectively with different number bases and with finite arithmetics.

Above all we must know that it is in the most rule-bound numerical activities that the machine is most error-prone and inefficient unless *we* remain firmly in the driving seat to direct and control its workings.

Exercises

1) What is $3*10^3 + 4*10^2 + 7*10^1 + 6*10^0$ worth in total?

2) What is the value of $|3*10^7 - 4*10^5|$?

3) Convert to base 16:

   a) 439_{10}
   b) $16,000_{10}$
   c) $32,160_{10}$
   d) $4,275_8$

4) Convert to base 10:

   a) $4AF_{16}$
   b) 110011101_2
   c) 43057_8
   d) $FFFF_{16}$

5) Evaluate:

   a) $349*10^{-3}$
   b) $0.0285*10^6$
   c) $349A*16^{-2}$
   d) $2.AF3*16^3$

6) Evaluate:

   a) $(11*4)_{mod12}$
   b) $(11 + 3 + 7 + 9)_{mod\ 12}$
   c) $(7*10)_{mod\ 11}$
   d) $((6*9) + (5*8) + (9*7))_{mod\ 11}$

6
Mistakes and Error

Objectives

After working through this chapter you should be able to:
- distinguish between mistakes and error;
- recognise how and where mistakes are likely to be made and understand elementary detection procedures;
- understand inherent and induced error, the absolute and relative size of error and how error increases in processing;
- identify error induction in the computer;
- calculate error by commonly used methods;
- recognise how error is propagated and take elementary precautions to minimise error.

6.1 Mistakes

6.1.1 One essential difference which we must always have in mind when adapting mathematical ideas and techniques for practical use in computing is that between MISTAKES and ERROR.

We have, in everyday speech, somehow managed to diminish the word 'error' by using it as a 'softer' criticism. Rather than simply saying, "You have made a mistake" we have persuaded ourselves that it is less offensive to say, "I think that there is a small error here."

Such attempts at being less brusque are fine in everyday social matters where politeness is important, sometimes to the point of masking the truth. We also often even refuse to admit, at least to anyone else, that we ourselves ever *make* mistakes!

6.1.2 In the much more disciplined world of computing, as in mathematics, it is important to realise that, although all human beings make mistakes and may, if they are sensible people, learn from them, *any single mistake could have been avoided*. It is also very important to learn to admit to our own mistakes and not try to 'bury' them!

6.1.3 All our systems should, as a matter of common sense and good elementary design, take account of this human fallibility. To ignore it in the hope that it will not affect *our* systems is simply foolish. We *must* assume that mistakes *will* always be made, try to help people detect and correct them and, last but perhaps most important of all, ensure that a system will not be put in jeopardy by any mistake which we might reasonably have foreseen.

We must also encourage the best possible training for the people who work with our systems, encourage them to seek help when a mistake *has* been made and not to be afraid to 'own up'. Considering the GIGO problem in computing, the role of the data-prep(aration) staff is often vastly under-rated and they are often under-trained. If a mistake has been made, at data-prep, data-input or in processing, the output *will be corrupted*, and, worst of all, this will not necessarily be apparent!

6.1.4 Mathematics can do nothing to prevent mistakes being made; it can, however, offer some elementary help in *detecting* mistakes *before* they can affect the integrity of the processing or cause degradation of the output.

Mistakes in processing will almost always be either as a result of imperfect program design or testing *or* some failure on the part of the operator. However the most vulnerable of all the elements of any system is the man/machine input interface.

6.1.5 Among the kinds of mistake which are commonly made in any data input phase are:

mislaying source documents;
mis-reading of data;
wrong data type;
wrong data range;
mis-keying data;
transposition of characters;
character drop-out;
bit drop-out in data transmission from remote input.

The list is by no means exhaustive but will serve to indicate which types of mistake may be detected by *some* mathematical process. We have talked earlier about validating data types and data ranges which means that well designed *and tested* data-input validation routines are elementary requirements.

6.1.6 Batch quantities and batch totals, counting and summing relevant numbers of items (for example money fields) in a set of documents are fairly obvious examples of 'ordinary' arithmetic being used to detect mistakes.

A little less orthodox is the idea of the 'hash' total. We can, for verification and validation processes *only*, add together any set of numeric fields, invoice dates for instance, or customer numbers, to generate check totals from ordinal numbers. Indeed these are so useful for detecting mistakes that it is often worth using purely *numeric* data types for such purposes even if alphanumerics have been used in the preceding manual system. Hash totals may equally well be used, internally, for *alphanumeric* fields of special importance.

Since all the letters and symbols are dealt with in the machine by numeric codes, we may total the codes rather than the symbols which they represent. All the character codes in, say, a sequence of characters in transmission or even of all the code of a program, may be added to form a 'checksum'.

The use of parity bits in checking for any corruption of a single character is almost too well-known to merit lengthy discussion. In ODD PARITY a 1 is placed in the 'leftmost' bit position of an 8-bit byte when there is an *odd* number of binary 1s in the remaining seven bit positions. If 'A', coded in ASCII 7 bit code as 100 0001 is represented in odd parity it becomes 0100 0001 because it *does not* have an odd number of binary 1s in the original 7-bit form. If it were corrupted into, say, 0100 1001 at some point, a parity check would reveal that something was wrong since with three 1s in the rightmost seven bits the PARITY BIT ought to be 1.

Clearly this can only be used, on an eight-bit bus or communications channel, if we need only seven bits to code all the characters which we use. Since today we often need the extended, eight-bit character set we may be restricted to the 'no parity' condition. Parity checks cannot then be used.

6.1.7   One other very common group of mistakes is those leading to copying in, to a data-prep form or the keyboard, ordinary, but especially *long*, numbers.

It is extremely easy, between eyes and fingers, to input say, 1058535 as 105535 or 1085535. Each of the three forms is the right data type, within a similar data range and hence unlikely to be detected by some standard checks. Such mistakes are, however, among the most frequent in any data-handling and transcription process.

If it is a number of some importance which is in frequent or permanent use, like a customer number, or even if it is used as an internal hash totalling process, we may use a standard arithmetic process for generating a hash total and appending it to the end of the number as a checksum.

With 1058535 we could add the numerals together. as $1 + 0 + 5 + 8 + 5 + 3 + 5$ to give a total of 27 which we could then tack on to the end of the number to make 105853527.

This, on the face of it, *would* show up a digit drop-out, *if* we knew where the original number 'stopped' and the checksum 'began'. We don't!

Furthermore, we have no *way* of knowing since the checksum may, in this type of example, be any size, almost at random. For instance the checksum for 23 would be 5; for 28 it would be 10; how would we know how big a checksum to expect or where to 'cut if off'?

6.1.8   This is a natural feature of any INFINITE arithmetic; the numbers just keep on growing. Only a FINITE arithmetic will give us numbers of fixed length, no matter how large the 'original' number. If we used MOD7 numbers, for example, the checksum, after adding all the digits, for 1058535 would be 6, for 23 it would be 5, for 28 it would be 3; always a single digit to append to the subject number and to strip off for checking.

Unfortunately, however, whilst this *would* detect any digit drop-out, (unless that digit were 0, which is *not* unimportant), it would *not* detect a transposition of digits, as in 1085535, because the checksum would be the same, and in consequence, so would the check digit.

It is to overcome this problem that we do some more unorthodox arithmetic by 'weighting' the columns before we hash total the digits. We do this by multiplying the numeral in each column by a *different* fixed factor. In ordinary commercial practice we keep it as simple as possible by giving 'weighting factors', that is to say multipliers, to the columns in the sequence 2* for the units, 3* for the tens, 4* for the hundreds, 5* for the thousands . . . . . and so on. In our example we should then have:

$$8*1 + 7*0 + 6*5 + 5*8 + 4*5 + 3*3 + 2*5 = 117_{10}$$
$$\text{and } 117_{10} = 5_{\text{MOD 7}}$$

If we now check our transposition error, where we had 1085535 our checksum would be $120_{10} = 1_{\text{MOD 7}}$ and we should know at once that the number had been incorrectly entered.

6.1.9   In practice the weightings may be kept secret, or even, where security is of exceptional importance, be changed at frequent intervals.

The check digit actually used is customarily the COMPLEMENT of the MODU-LUS as this leads to a considerable improvement in the simplicity of checking. Applying this to our first example, using MOD 7, for instance, the next to the last step would be to *subtract* the modular value of the checksum from the modulus itself, 7–5, to generate the checkdigit 2.

When we have taken this final step we can make 1* the weighting factor of the new 'units' column, containing the check digit, and recalculate the checksum without 'stripping off' the check digit. If all is well the new checksum should be 0, whatever the modulus.

Let us verify that with our number, 10585352 and weight, from right to left, by factors of 1, 2, 3, 4, 5, 6, 7, 8 respectively. You should arrive at a checksum of 119_{10} which, in MOD 7, equals 0 and shows that the number has probably not been corrupted. I use the word 'probably' because there are some transpositions which will not be detected. Can you suggest why?

Some 90% of such mistakes *will be detected* and that makes the technique vitally important to us.

6.1.10   Before leaving this topic we should look back to the previous chapter and our exploration of finite arithmetics. We saw that the prime number, 11, had many fewer repetitions in its number squares than did 12, when used as a modulus. For this reason we always use a prime number modulus, often MOD 11. What we need then, however, is another symbol, commonly X for 11 itself (with 0 for 10) if we are to keep to a single check digit.

There are advantages to using still larger numbers but the small increase in mistake-trapping is usually outweighed by the additional complications in processing. There is always a very difficult judgement to make as to when relatively small gains, in effectiveness or efficiency in one aspect of an application, are at the expense of slowing processing to an unacceptable degree. We can take it that MOD 11 is probably the commonest in commercial and industrial use.

6.2 Error in General

6.2.1   To the 'pure' mathematician numbers are clean, precise, even beautiful. In the real world, in which *we* have to function, pure numbers are of little importance and numbers are *usually* imprecise, approximate or may even be downright treacherous! Any real life number, 3, for instance, must always be accompanied by some label. We need to know "3 what; . . . . . dogs, metres, dollars, ...?"

The number is useless to us unless it is *describing some reality*. We talk then of *applied* mathematics, ie, applied to the description of real entities and solving real problems. Pure mathematicians often find this a very frustrating world.

6.2.2 What we also have to be concerned about is that our machine will only process the 'pure' numbers. We must strip off the labels *before* processing and ensure that we replace them afterwards because a number on its own makes no sense to people in the real world. This is a primary purpose of our variable names; they must be mapped on to the labels of our numbers.

6.2.3 Whenever we label a number, however, it is because it is describing a measurement on some SCALE and this is where the world of ERROR takes over from the neat and tidy world of pure mathematics. Every number which is a model of reality has, with very few exceptions, INHERENT ERROR. Only when we are using {natural numbers} for counting a small collection of entities, using the +ve integer number line as our scale, do we have numbers without this error, which is *not* artificial and *not* due to mistakes or carelessness. Nor can it be eliminated, no matter how concerned we may be with accuracy. It is an entirely inevitable attribute of the number in its scale.

6.2.4 There *are* situations in which a number will be rendered less precise because of something which *we* do, directly in calculations, indirectly in the way we write our programs *or because of the limitations of the computer itself*. We call the effects of these, by contrast, INDUCED ERROR, because they have been 'brought into' the situation, as it were from 'outside' the numbers themselves.

Error will grow during processing: unless we are extremely careful in our computer processing it may well grow out of control. Every inherent error may be looked on as a seed from which much greater error may grow and we call that process error PROPAGATION. There is, as you see, a considerable new vocabulary to be learned and, a little later, some new symbols.

6.2.5 First let us look in more detail at inherent error.

To take a rather crude illustration first of all we can see, in Figure 6.1, the effect of measuring a value, *V*, shown by the thick line at the top, on a scale of units where, in this instance, we can see that $2 \leqslant V \leqslant 3$, or *V* is somewhere between 2 and 3.

| 0 | 1 | 2 | 3 |

Figure 6.1 A number on a unit scale

If we estimate, by eye (see Figure 6.2) we can see that $2\frac{1}{2} \leqslant V \leqslant 3$, but V is still 'somewhere in-between' and we can never be quite sure just where.

6.2.6 There is always this element of uncertainty and it is in this that our concept of inherent error originates since any value lies in just such an area of vagueness and we can never be quite sure where.

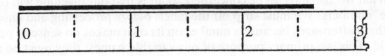

Figure 6.2 'Eyeing' the halves

6.2.7 No matter how much we may sub-divide our scale into smaller units there is still a region of uncertainty, one scale division in size.

Figure 6.3 gives an illustration of this and the consequent inevitable nature of inherent error.

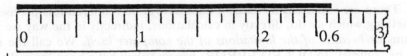

Figure 6.3 There is always uncertainty within one unit (or scale division)

6.2.8 When we state that a measurement has the value 3 we are saying that, on the scale used, its value is greater than $2\frac{1}{2}$ and less than $3\frac{1}{2}$ but *we don't know, and cannot know, just where, between those boundaries the 'true' value lies!*

To indicate this uncertainty we should, if we were being extra careful, write that the value is $3 \pm \frac{1}{2}$. Ultimately, no matter how fine the sub-divisions of our scale, the ERROR BOUND is always $\pm\frac{1}{2}$ of one of such sub-divisions.

6.2.9 In all cases such as these the numbers have a +ve and a –ve error component of approximately the same size. This means that the error bound is always symmetrical and bi-lateral, that is to say that it always has *two* parts of equal size. In our example the total error range, within the error bound, is 1 unit, but the largest amount by which any *actual value* may differ from the REPORTED VALUE, 3 (and ignoring the +/– symmetry) is $\frac{1}{2}$.

6.2.10 When we ignore the sign of a value and only concern ourselves with its magnitude we speak of the ABSOLUTE VALUE of that magnitude which we write as $|\frac{1}{2}|$ or $ABS(\frac{1}{2})$. We may find this expressed in symbols as $E_a = 0.5$.

6.2.11 Here we must take note of the ultimate case of the effect of E_a. The difference between any TWO values REPORTED as 3 is that one of them could be as small as 2.5, the other as large as 3.5. In other words, when we are comparing two or more values, reported as identical, the possible error is $2*E_a$!

This is of the greatest importance in computerised manufacture, or any other application of computing to any field of engineering. If a cylinder is bored to a diameter of 30mm and the piston is turned to the same reported (or nominal) size it could happen that the piston was 30.5mm in diameter and the cylinder 29.5mm in diameter! Not much hope of a good fit there. Absolute error, therefore, may present some unexpected difficulties if we are not on guard against such extremes.

6.2.12 In manufacturing industry it is assumed that sizes of any reported value may fall anywhere within the error bound and will be distributed statistically. Process control will often call for computerised statistical models to suit.

6.3 Relative Error

6.3.1 Clearly, if all measurements on the same scale have the same error bound and consequently the same absolute error the *effect* will differ according to the reported value. $E_a = \frac{1}{2}$ on a reported value of 3 will be comparatively large whilst $E_a = \frac{1}{2}$ on a reported value of 50 will be much less so. For this reason we find the absolute error useful mainly as a stepping stone towards the more meaningful RELATIVE ERROR, commonly written as E_r.

In this we express the absolute error *as a proportion of the reported value* and, as with many other proportions, we usually express it as a percentage. (We can also convert the error bound to a *relative* percentage to give us two possible forms.)

a) $100*(E_a/R)$, (R = the reported value) gives the *absolute* value of E_r, in the case of our reported value of 3, then, we have $100*(\frac{1}{2}/3) = 16.6\% = E_r$; *or* any *individual* value reported as 3 may be 16.6% larger or smaller than the reported value.

b) $100*$(error bound/R) = maximum possible relative error *between any TWO values reported as R*.

In our example once again, $100*(1/3) = 33.3\%$, is the maximum possible E_r between any two such values.

We most commonly use method a) but we must be aware that if we happen to forget the 'larger *or* smaller' proviso we could seriously underestimate E_r in some situations.

6.3.2 If we now compare our reported values of 3 and 50 using E_{r1} for the former, at 16.67% (corrected to two places of decimals) how does this compare?

$$100*(\tfrac{1}{2}/50) = 1\% = E_{r2}.$$

This makes it very clear that the *size* of any reported value is a very important factor in any consideration of its inherent error. But only up to a point where any practical 'fit' is concerned. A piston of 50.5 cm diameter will still *not* fit into a cylinder of 49.5 cm diameter!

6.4 Error Propagation

6.4.1 Let us, at this point, make our first investigation of ERROR PROPAGATION.

If we take reported values of 2 and 3 units and add them we have a reported result, 5. Taking account of inherent error we have, at the minimum,

$1\frac{1}{2} + 2\frac{1}{2} = 4$ and, at the maximum
$2\frac{1}{2} + 3\frac{1}{2} = 6$ and $E_a = 100*(1/5) = 20\%$

E_r on the 2 is 25% and on the 3 is 16.67%.

6.4.2 On adding two values the final, or accumulated, error is somewhere between those on the individual values. Can you, by taking a number of such examples, derive any rule for the accumulation or propagation of error under addition?

6.4.3 To examine the effects in multiplication we may use a visual model of the situation such as $A = L*B$. Here, in Figure 6.4, we see the white rectangle, whose reported sides are 2 and 3 units respectively.

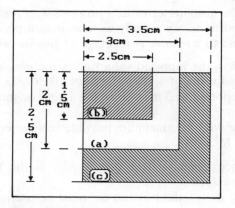

Figure 6.4 Multiplication explored as L*B

The reported area is 6 square units. The smallest possible rectangle to fit the 2*3 reported lengths is shown hatched and is $1\frac{1}{2} * 2\frac{1}{2}$ or $3\frac{3}{4}$ square units; the largest is $2\frac{1}{2} * 3\frac{1}{2}$ or $8\frac{3}{4}$ square units. The error bound is $-2\frac{1}{4}$ to $+2\frac{3}{4}$ square units and we see that INDUCED ERROR may not be symmetrical.

This, in the long run, is quite important but let us, for the moment take E_a to be half the error bound which makes the error $2\frac{1}{2}$ square units! Note that we still need the label for our absolute error although it disappears when we write E_r since that value is always a proportion.

In this case E_r is clearly 41.67% since the model $100*(E_a/R)$ produces that answer. In other words we have an apparent error of 41.67% as a result of multiplying quantities with E_r of 25% and 16.67% respectively.

If we take the larger 'side' of the error bound as the safe margin, E_a would be 2.75 rather than our 'nominal' 2.5 and E_r would be 45.83%. Clearly we can say that induced error in multiplication is found by summing, adding together, the contributory inherent errors, if the error bound is symmetrical.

If, as is often the case, the error bound is not symmetrical the induced error may be slightly more or slightly less than the nominal error.

6.4.4   Since, in computing, we often set up our models to perform whole sequences of multiplications let us take one step further and look at Figure 6.5, which shows three-stage multiplication by way of modelling a cube of sides 3 units long.

The reported value of the volume will be, of course, $3*3*3 = 27$ cubic units. The smallest, shown hatched in Figure 6.5, will be

$$2.5 * 2.5 * 2.5 = 15.625 \text{ cubic units}$$

with the largest $3.5 * 3.5 * 3.5 = 42.875$ cubic units.

The error bound is, in total 27.25 cubic units, which is *larger than the reported value itself!*

For any practical purpose the result is useless, although it is a dramatic illustration of the propagation of error in deceptively simple models. If we quantify it as E_r we find that it is again non-symmetrical but the average of the –ve and +ve elements is 13.625 and E_r is 50.46%; this is three times the inherent error in the length of a side.

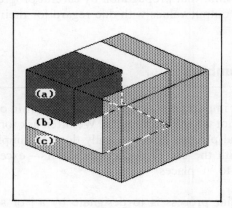

Figure 6.5   N*N*N in the form of a cube

6.4.5   You should try several examples of such additions and multiplications to be quite sure that you understand these effects. They are of the greatest possible importance to us.

6.4.6 If we look at subtraction, sticking to our reported 2 and 3, 3 – 2 can be, at the most 3.5 – 1.5 = 2; it may be, at the least, 2.5 – 2.5 = 0 giving $E_a = 1$ and $E_r = 100\%$. Try some other examples and see if you can arrive at a rule for error propagation in subtraction.

6.4.7 In division 3/2 has a maximum value of 3.5/1.5 = 2.3 recurring and a minimum of 1.

E_r is then 44.4%

You should try a number of other divisions, starting with the inverse, 2/3, and see if you can evolve a general rule for division.

6.4.8 Finally try $\dfrac{2*3*4}{5*6*7}$

    a) first multiplying all the top row, then all the bottom row and finally doing the division;

    b) second doing a multiplication followed by a division and so on;

    c) third 'cancelling' any possible pairs first and then solving the rest.

Which process causes the least error propagation and how does this affect program design?

6.4.9 The problem that we have in practice is threefold;

    a) to recognise how and where error occurs and is propagated;

    b) to control and limit that propagation by careful program design and testing;

    c) to present our output values with due regard for their error component.

6.5   Rounding Numbers

6.5.1 There are other important sources of induced error in our ordinary arithmetic processes, whether or not we use a computer. We often ask for values to be 'rounded to one or more decimal places'. This is usually done to simplify manual arithmetic and is less common with the use of pocket calculators – except that some of these offer options to round to 'n' places.

6.5.2 The most familiar process is to designate by a | or something similar, the column at which rounding is to take place, for example.

    a) 3.141 | 1592   b) 3.1415 | 92
    c) 4.743 | 5   or   d) 4.744 | 5 say,

and if we describe this as the 'cut' we may say, "If the numeral in the column immediately to the right of the cut is >5 add one to the numeral in the column immediately to the left of the cut; if the numeral to the right is <5 do nothing; if the numeral to the right is 5 then EITHER, if that to the left is odd do nothing, OR if that to the left is even add 1."

Perhaps we should pause momentarily here to note that when we round *up* we actually make a number larger than it was originally, if we 'do nothing' we are actually cutting off the remaining digits and making the number smaller. By taking the 'odds or evens' approach to the right hand 5 we are balancing out the ups and downs. The error bound induced is then ± 0.5 *of the value of the 'cut' column*. If we cut at the third place E_a = 0.0005 and when we look at any number which has been corrected to, say, 3 places of decimals, commonly written 3D, this is the error which has been induced into it.

To return to our examples, the 'new' numbers above are:

a′) 3.141;   b′) 3.1416;   c′) 4.743   d′) 4.745.

If we were simply shown these numbers, in ignorance of their *original size*, we would have to apply E_a = 0.0005 but here we *do* know what they were originally and we can compare the *actual* with the *rounded* value. If, for example, we look at that difference, and we find differences by subtraction, ignoring the sign if we need the ABSOLUTE DIFFERENCE, in the case of example a), above, |a – a′| is 0.0001592 and this is the E_a which we have induced by rounding to 3D.

The resulting error is *unilateral* and E_r is 0.49% which is quite small. (How long, however, would it remain small if it were involved in a sequence of multiplications, say?)

Notice that in this case the error bound is unilateral, one-sided. Try the others in that little group and arrive each time at E_r. Then see what the effect would be if we rounded to 2D.

6.5.3   This type of induced error is one which we can control relatively easily by careful consideration of the processes involved and by postponing all rounding to the last possible moment in our calculations. It is a very good argument for using the high precision variables in many commercial calculations; subject always to the cost in processing speed.

6.6   A Rounding Model

6.6.1   It occurs quite often that we, for various purposes but especially for the meaningful rounding of numbers *before* publication, need to include a rounding procedure within a program or a spreadsheet.

What do we need to do?

1) decide a 'cut' point
2) decide whether we should round up or down
3) achieve the rounding.

Here is a possible model for rounding to 2D:

```
read n
A ← INT(n*1000)
B ← INT(A/10)
IF (B–((INT(B/10))*10))/2 = 0
    OR IF A–(10*B) > 5
        THEN A ← (A+5)
    ENDIF
n ← (INT(A/10)/100)
write n
END
```

This is not necessarily a very good model but you should test it thoroughly and try modifying it for rounding to 3D for the value of the practice. It does show the decisions which we make in our minds when we are rounding numbers and it shows some of the ways in which we can 'tell' the computer where our cut comes and how to deal with the 'next column to the right, next column to the left' part of the decision-making.

The machine, of course, doesn't *know* right from left.

There may be, in the translator which you are using, 'invisible' sub-routines, called by a single command, such as 'CUT' or 'FIX', which will accomplish part or all of such a task. Even if they are present we must still understand what they do and how they do it if we are not to be blind when dealing with error propagation.

6.6.2   In a situation where the exact balancing of the 'rounding of the fives' is not too critical we can, in a spreadsheet for example, often use a single formula such as (INT((n*1000)+0.5))/1000) to round to 3D.

6.7   Significant Figures

6.7.1   A further source of induced error is the reduction of numbers to significant figures. The argument goes something like this:

"If we have a number of substantial size, say, 1058535792, the last few numerals, say the 5792, are of such small size compared to the overall size of the number that they are really insignificant. Indeed they may well be completely spurious as in stating that the population of Ruritania is 53475372 since no-one, no matter how accurate the counting process, could arrive at a number of such precision. Where such numbers give a false impression or where they are too cumbersome in our arithmetic, let us quote only those five figures, say, which give an adequate impression of size."

The result is that, to 5 significant figures, often written as 5 sig. figs., 1058535792 would become 1058500000 and 53475372 would be 53475000. It is clear that we have created or induced error in just the same way as does rounding to nD. Again the error will be unilateral and again it may not have a disastrous effect on the original number.

The revised value, however, must only be used in computations if the consequent error propagation has been foreseen and guarded against.

6.7.2 We can, of course, express real numbers to 'n' sig. figs.; in 2.0375612, just as in our population count, above, the last few digits are very small compared with the left hand 2. We could well express this as 2.0375, i.e. to 5 sig. figs. NOTE that any zeros essential to the true size of the number must be kept.

6.8 Truncation – A Procrustean Bed!

6.8.1 In ancient Greece a rather villainous man, Procrustes, would entertain guests right royally so long as they would consent to spend the night under his roof. They had, however, to agree to sleep on his 'guest bed'. Once there, alas, if they were too short to fit the bed they were stretched, forcibly, until they did; if they were too long the 'extra' bits were chopped off!

6.8.2 Within the computer we have yet another set of error-inducing situations and the first of these is register size. Any computer has a FINITE MAXIMUM SIZE of number which it can contain and numbers with digits in excess of this will be subject to induced error in two different ways.

By CHOP or CUT where any excess digits at the righthand end of a number, ie the LEAST SIGNIFICANT digits, will be arbitrarily discarded. This only induces 'normal' rounding or sig. figs. measures of error and can be accounted for comparatively easily.

6.9 Overflow Error

6.9.1 Much worse is OVERFLOW error, where it is the leftmost, MOST SIGNIFI-CANT, digits which are discarded. This, of course, may be quite catastrophic and is usually irrecoverable. Fortunately many writers of compilers and interpreters help us by delivering an OVERFLOW ERROR message but the cut may already have been made even though our processing has been halted. The prevention of, warnings about and recovery from, overflow errors is really very critical.

6.10 Changes of Number Base

6.10.1 A final example of internal error-induction is in the CONVERSION ERRORS induced when changing decimal fractions to bicimal form or {binary fractions} = {1/2, 1/4, 1/8, 1/16,........}. If we take any decimal fraction which is a sum of any of the elements of this set we have no problem. 0.75 for example, is fine because it can be expressed also as 1/2 + 1/4 but let us look at 0.7, which would, on the face of it, seem a simpler value to transform. For convenience let us arrange for the conversion to follow a similar form to the denary/binary conversion we looked at earlier but we will put the binary result on the top line rather than at the side, as in 'long' division. We will, at each level divide by the appropriate binary fraction in the sequence but, for consistency of number operations, express these in decimal form.

6.10.2　First, how many halves in 0.7; how many quarters in the residue; how many eighths; how many sixteenths . . and so on.

```
              0.10110011      BASE TWO FRACTION
       0.5 | 0.7
           | 0.5
      0.25 | 0.2
     0.125 | 0.200
           | 0.125
    0.0625 | 0.0750
           | 0.625
   0.03125 | 0.01250
  0.015625 | 0.0125000
 0.0078125 | 0.0078125
0.00390625 | 0.00468750
0.00195325 | 0.00390625      and so on.......
```

6.10.3　This would be comparatively straightforward to program and you should try a pseudo-code version, but there is an alternative model which is perhaps not so easy to understand but is much simpler to set out for pencil and paper working. In this we 'fix' the columns firmly on the decimal point and perform repeated multiplications by 2 of the digits to the *right* of the decimal point.

As the first step

$$0.7*2 = 1.4$$

and the '1' is the first digit of our answer.

Further digits will be all the 1s and 0s as we go along. The second step, multiplying only the digit to the right of the decimal point, will be:

0.7
$$1.4*2 = 0.8$$

and the complete process, to 10 'places of bicimals' will be:

```
Read down    ↓
            |.7  *2
         1 |.4  *2
         0 |.8  *2
         1 |.6  *2
         1 |.2  *2
         0 |.4  *2
         0 |.8  *2
         1 |.6  *2
         1 |.2  *2
         0 |.4  *2
         0 |.8  *2
```

Take it several stages further and see if the '0011' sequence continues, or whether the bicimal fraction 'works out exactly'.

If there were room only for 6 of the bits in the above answer, which is, of course, the same by either method, can you see how to evaluate the consequent error?

As always you should try this for a variety of numbers to be sure that you have grasped the routine.

6.10.4 It is not difficult to see that there will be many such finite decimal values which generate irrationals in the binary system and will be truncated when the register is filled. This used to be something of a problem in the early pocket calculators which baffled many people when 3*(10/3) did *not* return the expected answer! The programmers of those processors now look after us a little better.

6.11 The Control of Error

6.11.1 As an example of the need for the control of error in processing let us look at the model of the volume of a cylindrical cone:

$$V = \frac{1}{3} \pi r^2 h \text{ where } r = \text{radius and } h = \text{height.}$$

If we assign some test values:

$$r = 2.2 \text{ cm}$$
$$h = 12.5 \text{ cm}$$

and take π as being 3.141 we have:

$$V = 0.3 * 3.141 * 2.2 * 12.5$$

and we must next look at the error in each number or process in order to assess the extent of error propagation and whether we can minimise it.

We must also do this, as a routine part of our modelling/programming so that we can establish (and this is really extremely important) *what degree of accuracy we may legitimately claim when we publish our answer*. We far too often publish results to many more places of decimals, or to many more significant figures that can possibly be justified by the accuracy of our numbers or of the processing! Such answers are not merely pretentious but are often dangerously misleading.

6.11.2　Let us subject our model to such a scrutiny:

The errors present are:

n	E_a	$E_r\%$	
0.3	0.05	1.67	induced by rounding
3.141	0.0005	0.02	induced by rounding
2.2	0.05	2.3	inherent – but more will be induced by conversion. (below)
12.5	0.05	0.4	inherent.

This totals, since all elements are to be multiplied, 4.4%. This means that in a calculated value of 25.91325 we have an error of ±1.14 cubic centimetres. How meaningful are all those decimal places in our uncorrected answer?

6.11.3　The next step is to see if we can, by planning our calculation, reduce that error. What happens if, for example, we work in integers and do all our re-conversion at the last stage? This would imply also dividing by 3 at a late stage rather than multiplying by $\frac{1}{3}$.

Try this and compare the two results. This is just an illustration of how we must examine all our operations on models to exercise control over error propagation and to publish only reliable results.

6.12　Conclusion

Given all these ways in which error is present in numbers, is induced in our operations and propagated by our processing, it is small wonder that any computer person who is without a firm grasp of these matters may experience some very severe shocks and allow some very strange results to be published.

Exercises

1) List five types of mistake which may occur during the data prep and data entry stages of the processing cycle.

2) Give two examples each of the appropriate use of hash and batch totals.

3) Describe the principle of a parity check and how even and odd parity differ.

4) Using mod 7 and weightings of 2, 3, 4, 5 . . . . . for the units, tens, hundreds columns, and so on, append appropriate check digits to:
   a) 439
   b) 27946
   c) 30000
   d) 1058535
   e) 17493726

5) Using the same procedure as in Question 4 determine which of the following numbers, which include check digits, are valid:
   a) 23756
   b) 1794
   c) 200004
   d) 1743562
   e) 10545722

6) What is the error bound of values reported as:
   a) 6km;
   b) 3.5cm;
   c) 1.05 grammes;
   d) 350cm?

7) Calculate the maximum relative error on:
   a) 3.5
   b) 13.5
   c) 130.5
   d) 1300.5

8) Calculate (3.5*13.5*1.5) / (7.0*4.5*40.5) in such ways as will:
   a) maximise;
   b) minimise, the accumulated error.

9) What is the effect on error if 3.5*7.25 is:
   a) calculated to integer precision;
   b) cut to 1D;
   c) rounded to 1D;
   d) calculated to 3 sig figs.;
   e) worked out 'in full', with no cut or rounding?

10) If $V = \dfrac{4}{3}\,\pi r^3$ is to be calculated to 3D what precautions, if any, should we take about the values used in the calculation and the sequence of calculations in our model?

7
Storing and Manipulating Numbers

Objectives

After working through this chapter you should be able to:
- understand the way numbers are processed in the ALU;
- see how the architecture of the machine dictates the storage and processing methods;
- use different forms of coding for numbers;
- perform 'computer' addition, subtraction and multiplication;
- deal with floating point numbers;
- understand how error is induced in the ALU.

7.1 Why Should We Worry?

7.1.1 When first we learn to do pencil and paper arithmetic we quickly find that if we don't keep numbers in their proper columns our results soon become erratic.

Sometimes we shall be lucky and keep the true values, on other occasions we shall be misled and make mistakes.

When we enter numbers into the machine we simply assume that they *will* always be kept in their correct relationships with one another. How do we know? We have already seen that error is often induced by the machine and propagated by our procedures. How do we know exactly what to expect?

7.1.2 We can assume that the machine will do its sums *reliably*, that if we tell it to subtract it will do so and will always do so in the same way. We can by no means take it for granted that it will do our computations *accurately*! Here is reason enough for computer staff at all levels to have a good grasp of what actually goes on within the ALU.

7.1.3 Some students of computing will also go on to learn to program in assembly language or machine code, the so-called 'low level languages'. These, whilst being much more time consuming to write, are fast in execution and economical of storage space. To write such code we need a very intimate understanding of activities within the machine.

7.2 Some Preliminaries

7.2.1 First we need to link our ordinary arithmetic to what goes on in the machine.

One of our problems when reading computer manuals is that they often speak of aspects or techniques of mathematics which we may never really have understood or which many of us have simply done mechanically, as we were taught.

Let us go back to some of these basic activities and re-label or re-learn what is necessary.

7.2.2 We have seen in earlier chapters that arithmetic, which is all the machine can do with numbers, obeys certain laws. Among the most important of these, for our present purposes, are the laws of INVERSES.

These have the effect that, for *any* calculation, we can achieve the same end by *performing the inverse operation on the inverse of the value*. Rather than *subtract* 3 from 5 we can *add* –3 to 5.

A second point to remember is that multiplication is really only REPEATED ADDITION and division is no more than REPEATED SUBTRACTION. The simplest representation of 3 * 5 is 5+5+5 (*or* 3+3+3+3+3) but 3*5 is much shorter. In the same way we could write 12/3 as:

12–3; 9–3; 6–3; 3–3

and say that it took 4 subtractions to account for all the 12, or 'exhaust' it.

A primitive pseudo-code model of division could be:

```
Accept M,N
Q ← 0
DO UNTIL M < N
    M ← M–N
    Q ← Q+1
END DO
Display Q, "remainder", M
END
```

7.2.3 Arithmetic also has a *language* of its own which has arisen from the fact that any calculation, no matter how simple, must have four elements:

a)   an indicator of what *operation* is involved;
b)   a number to be *operated on*;
c)   a second number showing the *'scale' of the operation* so to speak;
d)   somewhere, in close proximity with a) to c), to *display the answer*.

3 + 5, for example, actually says something like, "to a starting quantity, three, *add*, (or count on) a further five, (or *increase* the 3 by 5)." Unless we also say ". . and tell me (or write down) what the total is" we shall be none the wiser.

The computer actually reflects this very accurately since if we merely enter, on the keyboard, 3+5, we shall never *know* if our instructions have been carried out! We must *also* enter something like 'display 3+5' in order to see the answer. Pocket calculators, on the other hand, are generally programmed to display the answer when we press the '=' button. We can deduce that this button performs a function similar to the 'display' command.

On paper we show the 'place for the answer' when we set out a sum like:

$$3+$$
$$5$$
$$\overline{}$$

by drawing the line under the lower number. If we set out the sum in equation style, ie as $3 + 5 =$ we know that we are expected to put the answer immediately to the right of the '=' sign. We have learned similar 'drills' for setting out all our pencil and paper calculations, to the extent that when we meet a new kind of calculation one of our first questions is "How do I set it out?"

7.3 Conventions

7.3.1 In mathematics we already have names for the *operations* and symbols to serve as shorthand for those names. We have 'multiply' to indicate repeated addition, for example, with *, ×, or . as the corresponding symbol. Much less familiar are the very ancient words for the three numeric parts of each of the four common kinds of calculation.

7.3.2 Since the operations are different the three words used in each case are also different:

For addition the first number is the	ADDEND
the 'second' and subsequent numbers is/are	AUGEND(S)
the result of the computation is the	SUM or TOTAL.

For subtraction the corresponding parts are	MINUEND
and that which is to be subtracted is the	SUBTRAHEND
with the result of the operation, the	DIFFERENCE.

Multiplication sees the rather clumsy	MULTIPLICAND
operated upon by the	MULTIPLIER
and producing, as a result, a	PRODUCT.

For division, which we set out differently	DIVISOR $)$ DIVIDEND
in pencil and paper calculations, we have,	QUOTIENT.

Getting to know these is of no great importance in itself but they *do* allow us to use the one word rather than writing, each time, something like 'the number we are subtracting from.'

They are of value while we are investigating our present topic but, happily, do not usually form part of everyday speech!

What we have learned in our schooldays of the 'setting out' of calculations, has included all these ideas but we have rarely learned them *as a set of fixed principles*. What usually happens is that we learn each 'drill' separately and rarely connect them into any coherent pattern. This often means that if we momentarily forget how to set out a particular problem we are defeated.

7.3.3   Even so we are much better at it than the computer which is actually incapable (with large and costly exceptions) of 'learning' anything. Much less can it learn and store different procedures for different kinds of sum.

Input devices do not even allow for such setting out. Indeed the machine will not even recognise the commas and/or spaces which *we* include in large numbers to make them easier to read. Where we might write 1,645,327 or 1 645 327, or some other notation which is familiar to us, the machine simply will not accept it.

We are left with a very rigid system which essentially recognises only the decimal point within a number and where all operations are conducted within the *same* framework, or 'setting out'.

7.3.4   If you have ever seen a mechanical calculator from the days before electronic machines took over you will have seen that they had three 'registers' in which the numbers involved in the calculations were placed or stored. This is shown in Figure 7.10.

These were very similar to rows of pigeon-holes which kept the numerals in their appropriate columns and kept them in their proper 'rows'.

One register held the number to be operated on, a second the augend, subtrahend, divisor or multiplier, whilst the third, generally called the ACCUMULATOR, contained the result. There was also some device for setting the decimal point to indicate which of the columns contained the 'Units'.

Since these were their only high-powered computational machines it was perhaps natural that the mathematicians who devised the earliest computer programming systems, including the control programs, should have used the 'register' ideas in the computer.

They did make some changes since we commonly have at least six registers and they can be 'doubled up' for some operations. Nevertheless, in the ALU, it is still essentially three sets of pigeonholes in the same relationship as the three 'rows' of a pencil and paper sum.

7.3.5   The computer, like the calculating machine described above, does differ from our ordinary arithmetic, in truncating numbers for instance, since *its registers are of fixed length*. On *paper* we can keep adding columns to accommodate any size of number we choose.

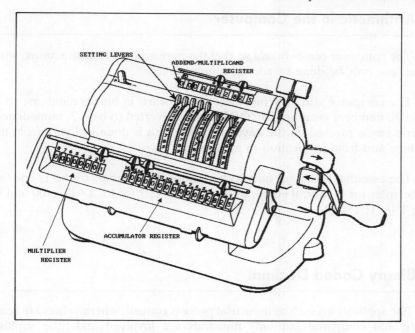

Figure 7.1   A mechanical calculator c. 1960

7.4   Less Familiar Arithmetic

7.4.1   Numbers have COMPLEMENTS, just like sets and just as with sets they 'complete a whole'. For numbers in our ordinary base 10 sequence the complement is what is needed to make up the whole of the appropriate 'power of ten' which 'contains' the number concerned. For example 7′, (the complement of 7), = 3 because 7+3=10; 437′ = 563 since 437+563=1000. We describe these as the 'tens complements' of the numbers concerned.

Just as we can subtract a number by adding its inverse we can also do so by adding its tens complement and finally subtracting the 'scale' value.

7.4.2   For example: 5 – 3 = (5 + 7) – 10

$$800 – 437 = (500 + 563) – 1000$$

Check these, and then write several others, to make sure that they work and that you can operate with complements. Can you see that in the first example all that we are really doing is taking advantage of the fact that –3 = +7 – 10?

In our ordinary arithmetic this is little more than an interesting oddity although it helped speed up calculations with the hand calculator of days gone by. It does, however, have great importance in the operations within the computer.

7.5   Arithmetic in the Computer

7.5.1   The computer *can only add* so that the performing of substractions, within the machine, can only be done by adding complements.

7.5.2   The computer also only functions, at its heart, in binary numbers, so that all our base 10 numbers must be ENCODED, or converted to base 2, immediately after a numeric key is pressed on the keyboard. How this is done will vary from machine to machine and from application to application.

7.5.3   The essential is that all such codes reduce the numbers to the 1s and 0s which obey the rules for binary: $0+0=0$; $0+1=1$; $1+0=1$; $1+=10$; $1+11=100$ and $0*0=0$; $0*1=0$; $1*0=1$; $1*1=1$.

7.6   Binary Coded Decimal

7.6.1   For applications such as industrial process control, where values are constantly recorded and compared but only numerals are involved and little sophisticated computing is necessary, it is still quite common for BINARY CODED DECIMAL to be used, usually abbreviated to BCD.

7.6.2   In this coding system each numeral is given a four-bit binary code to correspond with the list of ordinary counting numbers from 0 to 9. In this system each 'nibble', or half byte, of four BITS is effectively kept separate, even if moved around the system in pairs as 8-BIT BYTES.

The codes are:

$$1_{10} \quad 0001_2$$
$$2 \quad 0010$$
$$3 \quad 0011$$
$$4 \quad 0100$$
$$5 \quad 0101$$
$$6 \quad 0110$$
$$7 \quad 0111$$
$$8 \quad 1000$$
$$9 \quad 1001$$

leaving 6 'spare' BITS which do *not* represent numerals but which can be used for other purposes if needed. For example, it is possible to use three of them to represent the plus, minus and decimal point signs.

This is 'straight' BCD although more elaborate forms, including 6-bit and 8-bit versions, have been developed for more complex tasks.

Arithmetic with BCD is quite straightforward since the digits are kept in their nibbles and aligned in the registers just like ordinary numbers.

7.6.3 473 + 426, for example, would be

    0100 0111 0011   in the addend register.
    0100 0010 0110   in the augend register
    ─────────────────
    1000 1001 1001   in the accumulator.

If we translate the result back into denary we see that $1000_2 = 8_{10}$; $1001_2 = 9_{10}$ and $1001_2 = 9_{10}$ or 899.

This example does not involve any 'carries', ie no total, for any individual column, >1001, or 9_{10}.

7.6.4 Let us see what happens when we *do* have carries.
  597 + 458 becomes

    0101 1001 0111   addend
    0100 0101 1000   augend
    ─────────────────
          1111   partial sum

7.6.5 Notice how the partial sums, intermediate results, build up in the accumulator. Notice too that the sum of the units column, the rightmost column of nibbles, = 1111 which *is not* a 'legal' code in BCD where the largest possible numeral is 1001.

We must then *restore* the result to the legal maximum by *adding* 0110 to 1111, giving 10101 which we must enter and 'carry' 1 to the next column, to give:

    0101 1001 0111   addend
    0100 0101 1000   augend
    ─────────────────
                0101   partial sum

          0001   carries

which in turn gives the next column as 1001 + 0101 + 0110 = 0101 with 0001 as the carry so that, at the end of this stage we see:

    0101 1001 0111   addend
    0100 0101 1000   augend
    ─────────────────
          0101 0101   partial sum
    ─────────────────
    0001 ̶0̶0̶0̶1̶   carries

When we add the final column we have $0101 + 0100 + 1011 = 1\ 0000$ to give, as a final result,

	0101	1001	0111	addend
	0100	0101	1000	augend
0001	0000	0101	0101	accumulated sum
~~0001~~	~~0001~~	~~0001~~		carries

which we can translate back into 1055_{10}.

7.6.6 The leading zeros on the 'carries' are actually redundant. I have included them at this stage only to make clear that it is the *binary* one which is carried and that it is 0001 in its 4-bit, BCD form.

As always you should practise such BCD additions until you are confident with them.

7.7   More Sophisticated Codes

7.7.1   If we are dealing *only* with numbers this is a very simple procedure and could be handled by a fairly elementary processor.

It is too slow and takes up too much space for more general use.

7.7.2   Modern information processing demands much more of our systems than merely dealing with numbers and the American Standard Code for Information Interchange (ASCII) developed from the very old Baudot code, or better still the 8-bit EXTENDED ASCII, provides for processing far more than simple numerals.

7.7.3   IBM, which has always preferred to develop its own systems rather than adopt existing standards, took the extended BCD concept from 4- to 6- to 8-bit and eventually into its Extended Binary-Coded Decimal Interchange Code, EBCDIC, still largely restricted to IBM machines.

Both ASCII and EBCDIC were developed for handling verbal as well as numeric data but most of the complexities of handling text need not concern us greatly in this book.

7.8   Capacity and Register Size

7.8.1   The architecture of most modern computers, the whole of its internal communications, processing, control and storage, is based upon the 'word-size' of the system. Very small computers have 8-bit processors, medium-power machines have

16-bit architecture and high-power machines have 32-bit systems. Note that in this case the power of the machine is related to speed and capacity, not its electrical consumption.

7.8.2 From our point of view this means that each of the registers holds 8, 16 or 32 bits respectively. For commercial and industrial purposes 8-bit machines are rather slow and limited in memory size, although I have known all the information processing for a medium-sized engineering company to be handled on two 8-bit machines.

In principle a set of 32-bit registers functions in exactly the same way as do 16-bit registers. For the sake of simplicity I propose to concentrate on the medium sized, 16 bit, architecture but to take a brief look at integer arithmetic in 8 bit registers. The process is easily extended to 16- to 32-bit registers.

7.8.3 At the simplest, we can have, in an 8 bit register, any number from 0000 0001 to 1111 1111 or from 1_{10} to 255_{10}. Let us look first at integer addition, taking $129 + 89$ as an example, and then at integer subtraction using $129 - 89$.

7.8.4 a) addition
$$10000001 +$$
$$01011001$$
$$11011010_2 = 128+64+16+8+2 = 218_{10}$$

b) subtraction
$$10000001 -$$
$$01011001$$
$$00101000_2 = 32 + 8 = 40_{10}$$

This, however, as with any subtraction involves 'borrowing' from the next column to the left when we must subtract a larger from a smaller numeral in any column. All is well until we come to the 2^3 column when we must 'borrow' from the 2^4 column so that we can subtract the 1 from 10 instead of from 0. This gives us, in effect:

$$10$$
$$0111\cancel{0}001 -$$
$$01011001$$
$$00101000$$

and we can see the effect of 'borrowing one' from the leftmost 1000, now reduced to 0111. In this instance the remaining digits can be subtracted without further borrowing so we need complicate matters no more.

7.8.5 The computer, however, is not equipped to 'borrow and pay back' in order to perform a subtraction; it *can* only add.

The way best suited to the machine is to subtract by finding the complement of the subtrahend and adding. In binary arithmetic this is fairly straightforward.

7.8.6   We find first the 'ones complement' (ie, what we must add to make the number up to all ones):

> 01011001   by inverting each digit;
> 10100110   and convert to 'twos complement;
> 10100111   by adding 1 to the rightmost column.

If we remember our base 10 complement example we then *added the two numbers* and finally *subtracted* the power of the base group of which we had found the complement.

> 10000001   the original addend
> 10100111   the twos complement of 89_{10}
> 1   00101000   the sum of the two.

Note that a 1 has carried into the 2^8 column and this is precisely the size of the base 2 quantity of which we found the complement of our subtrahend. Do you remember, from our base ten example earlier, that our final step was to *subtract* this amount before publishing the result?

7.8.7   In most computer processors we do not need to take this final step because this type of overflow will simply be lost (Procrustes again!) so doing the last step automatically.

7.8.8   In some processors the whole operation is conducted in the ones complement by a slightly modified process.

Let us see what happens if we evaluate B – A, using a different pair of numbers. A=10001110 and B=11110000:

> 11110000   + = B
> 01110001   + = ones complement of A
> 1   01100001   + = sum – with 'carry 1'
> ↓ →→→   1   + = 'end-around-carry'
> 01100010   + = final result.

You should convert A and B to base 10, perform the subtraction, and show that the result is true.

7.8.9   When we define variables as INTEGERS it is quite common for any negative values, ie those being entered following a minus sign, to be stored and operated on in complement form which may be ones complement or twos complement depending on the processor used. Fortunately, we do not need to use such arithmetic in high-level language programming but at the machine code or assembler level we need to be competent in its use.

7.9   Multiplication

7.9.1   We can also perform a straight binary multiplication using nothing but addition and 'place shift'. Whenever, in a column system, we move a number one place to the left we actually multiply it by the number base in use. In binary arithmetic we can take 11*13, for example, and follow this process in the registers:

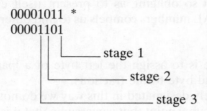

```
00001011 *
00001101
              ┌───── stage 1
            ┌─┴─────── stage 2
          ┌─┴───────── stage 3
```

Any multiplication by 1 leaves a number unchanged so if we perform the units multiply, followed by the 2^2 multiply, followed by the 2^3 multiply we see:

```
00001011 *
00001101
00001011 Stage 1
00101100 Stage 2
00110111
01011000 Stage 3
10001111
```

7.9.2   As always you should check this result by converting the base 10 calculation but note particularly that the place shifting has used up all the available places and overflow is imminent – with unpredictable results!

It is even worse if we have used the leftmost bit as a sign bit, which is effectively what happens with the twos complements numbers.

A special 'flag' will be needed to keep track, in the machine, of any corruption of bit 8 by overflow in that case.

7.10   More Powerful Arithmetic

7.10.1   Altogether we are seeing too many limitations to 8-bit binary manipulations.

7.10.2   Within an 8-bit machine the largest integer we could store in this way would be   01111111
and the smallest 10000000 since this is the smallest possible value in twos complement form.

The largest possible value then is +255 and the smallest –256; not much use for serious arithmetic!

7.10.3   The remedy is to use 8-bit registers in 'pairs' for all arithmetic operations, giving us 16 bits for addend, augend and accumulator registers and largest and smallest values of 0111111111111111, or +32767, and 1000000000000000, –32768 respectively.

By the same process a 16-bit machine, with paired registers, can hold +2,147,483,647 and –2,147,483,648 in integer form, which allows us to deal with quite large numbers.

7.10.4   Life, however, is not so obliging as to present itself entirely in nice neat integers and the world of REAL numbers compels us to find other means of allocating our registers.

7.10.5   The simplest of these is to assign the left byte of a 'paired register' to the integer part and the right-hand byte to the fractions.

So long as they are separately designated in this way we do not need to use up any space in the register for the bicimal point, but can assume that it is always between the two halves of the paired register. We are, so to speak, setting an invisible boundary between the two halves of the register as, for example,

$$10011011 \mid 10100000$$

which would translate as 155.625_{10}. The range of values available to us is still not large since the righthand byte is available *only for the fraction parts of reals* and it cannot now be used to enlarge the overall range of number values which we may use. It is, in this way, rather wasteful for much of the time.

7.10.6   In this notation the $\mid$ is not needed within the machine but we may need to introduce at this stage the concepts of 'most significant' and 'least significant' in order to clarify just how such things as an 'implied' decimal or bicimal point can be identified without using extra space.

7.10.7   We saw earlier that the word 'significant', in number work, refers to indications of relative size.

In this context the leftmost bit of an eight-bit byte refers to the largest component of the number represented by the byte as a whole. We describe it as the Most Significant Bit, abbreviated to MSB. The right-most bit is the *least significant*, or LSB.

7.10.8   In our paired registers, as they are used in the particular arrangement which we are discussing at present, the lefthand *byte* is also the most significant byte, since the righthand byte only contains the relatively small fraction part of the real number.

However, that byte too has its most- and least-significant *bits*; also the left- and right-most respectively. According to this we can say that, "The implied decimal point is between the LSB of the integer byte and the MSB of the fractions byte of the register."

We find such descriptions in quite common use in computer manuals but, once again, they are of fundamental importance when we work at machine-code, or assembler levels, using first and second generation languages respectively.

7.10.9 The straight-binary, split-register organisation of our numbers is another of the simpler devices since place-value is held by the implied point and is relatively fast in operation. We can, and do in many calculators, indicate different 'splits' for different purposes.

If, for example, we are working with a four byte register, we could split it so that 3 bytes served the integers and 1 byte the fractions. The smallest fraction we could then write would be

$$0.00000001 \text{ or } .008_{10} \text{ correct to 3D},$$

with a whole number part as large as 8,338,607.

7.10.10 If we split the registers in the other possible arrangements, 2:2, (ie two bytes for integer and two bytes for fraction parts), 1:3 or 0:4 we should have corresponding changes of balance between the range of sizes of integer and of fractional parts of reals. I have one electronic calculator, for example, which allows me to choose whether to work with all values CUT at 2 places of decimals without rounding; to calculate to 3 places and round to 2D; or to allow the decimal point to 'float'. This implies different arrangements of register split as outlined above.

7.10.11 The straight binary split register system is, however, both limited and rather wasteful of space. What it *does* give us on the profit side of the account, so to speak, is a pair of very valuable new ideas; the *split register* and *the implied point*.

Both are at the heart of the final and most powerful way of using the ALU which we are now going to look at.

7.11 Floating Point Real Numbers

7.11.1 Ideally our system should allow us to make the best possible use of all the space in the registers by showing as many significant figures as possible, whether the numbers concerned are large or small. Once again, the mathematicians and scientists had been using just such a system of calculating for many generations before the development of the electronic computer. We have met this 'scientific' notation briefly when discussing the column system in an earlier chapter.

7.11.2 Scientists, whose calculations quite commonly include both very large and very small quantities, learned to take advantage of two principles of number systems. First, it is easier to calculate if all the numbers we use are of roughly similar (ideally of *identical*) size in the 'registers', even on paper. Secondly, that the relative *size* of a number is indicated by the power of 10 of the most significant digit. The quantity 5351_{10}, for example, can be expressed just as readily as $5.351 *10 *10 *10$ and this in turn can be written as $5.351 *10^3$.

7.11.3 Here we meet a new notation. The 3 in 10^3 is known as the (base 10) EXPONENT and as we saw earlier such exponents may be +ve or –ve.

We write the complete 'scaling factor', $*10^3$, (which indicates the true size of our 5.351), as E+10 and the full 'EXPONENTIAL NOTATION' of 5351 becomes 5.351E+3.

7.11.4 For fractional values, for example 0.005351, which contains exactly the same numerals in the same sequence, we write 5.351E-3 which represents $5.351 *10^{-3}$.

Clearly we need a fair amount of practice to become fluent in writing numbers in this notation but it does unlock a powerful toolbox for making the very best use of our ALU in dealing with REAL NUMBERS.

7.11.5 If we partition our registers appropriately we can use the leftmost byte to contain information about the exponent and the remaining byte(s) to hold the sequence of numerals which complete the number. This latter part we call by the strange and rather exotic word 'MANTISSA'.

7.11.6 There are, then THREE PARTS to numbers held in EXPONENTIAL NOTATION: the EXPONENT, with –ve exponents stored in twos complement form: the MANTISSA; and the SIGN, (+ or –), of the entire number. We must note that it is equally possible to have –0.005351 or +0.005351 so that we may have a *negative number with a negative exponent*.

7.11.7 The application of these ideas to computing shows one important change from the pure 'scientific notation'. If we used numbers in the form 5.351E+3 we should need not three but *four* subdivisions in our registers, one for the integer, one of the fraction, one for the exponent and one for its sign.

7.11.8 We actually standardise exponential notation to *exclude* any integer part by writing 0.5351E+4 so that we can arrange our registers on a standard pattern of:

Sign (0 = + ve, 1= − ve) Exponent from 000001 to 1000000

Mantissa

1   0000111   0011011100011 . . . . . . to the limit of the register.

7.11.9 The arrangement here, for example, with seven bits for the exponent would cover the range of numbers from nE+63 to nE−64 with the 'n' indicating the mantissa. The leftmost bit of the exponent is set to 1 but is *not* part of the value and must be 'subtracted' at the final stage.

7.11.10 Using a 16 bit machine, with registers paired to give 32 bits in all and partitioned to allocate one byte for sign and exponent and 24 bits to the mantissa we can express and manipulate numbers in the range ± 16.777,215 E±55 and including all the fractional values in between.

Try to determine the maximum size of number which could be held in a sixteen bit register using one byte for exponent (including the sign bit) and one byte for mantissa.

7.12   Operations with FP Reals

7.12.1   There is a very large bonus for using this notation; we can do multiplications and divisions using a short cut. According to the associative and distributive laws we could write:

$$2000 * 30000 \text{ as } (2 * 3 * 1000 * 10000)$$
$$\text{or } (2 * 3 * 10^3 * 10^4)$$
$$\text{but } 1\,000 * 10\,000 = 10\,000\,000 \text{ or } 10^7.$$

7.12.2   In other words we can multiply exponential numbers by *adding* their powers! Our result is 6E+7. Try some others and then try some divisions. Does it follow that we can divide by subtracting powers?

7.12.3   It is a long-standing litany in conventional mathematics that:

Plus times Plus gives Plus
Plus times Minus gives Minus
Minus times Minus gives Plus
Minus times Plus gives Minus.

We, in our leftmost bit, are now using 0 to indicate Plus, a positive number, and 1 to indicate Minus, a negative number. In binary $1 + 1 = 0$ (carry one but overflow) and this maps onto Minus * Minus = Plus.

7.12.4   Do the others also map on to the 1s and 0s in the same way? What does this tell us about operations on the leftmost byte in multiplication and division within the computer when exponential notation is used?

7.12.5   Unfortunately there is a snag when we come to *add* numbers written in exponential notation. In the standardised form of mantissa and exponent the sum 0.4375E+3 + 0.257E+5 presents problems because we are not adding like to like, the numbers are of quite different scales or magnitudes. If both numbers were to E+5 or both to E+3 all would be well so we must convert one number to the same scale as the other in order to perform the addition (or subtraction by complementary addition). If we multiply the 0.257 by 100 and express it as 25.7E+3 we should, once again, need a separate register for the integer parts. This would defeat the whole system which we have developed. If, however, we express the 0.4375E+3 as 0.004375E+5 we can, using the distributive law, write our calculation as $(0.004375 + 0.257)*10^5$ or 0.261375E+5.

7.12.6   As always you should convert the original numbers from exponential to our ordinary notation, add them, and confirm that their total is truly 26137.5. Further practice of additions and subtractions of numbers, such as these, of different exponents and hence of different scales, will help you grasp the idea more firmly.

7.13 'Normalised' Numbers

7.13.1 We call this system of expressing numbers as 'standard' decimal fractions without integer parts but with the sign and exponent in the leftmost byte when converted into binary, the NORMALISED FLOATING-POINT SYSTEM of arithmetic. Ordinarily we assume, in all our mathematics, that a number without a sign 'in front' of it is a positive number and only negative numbers are signed when we write them. Values ≥ 10 will not usually have the sign of E given so that 4715, for instance will normally be written 0.4715 E4 although individual computer software may cause the leading zero and/or the space between the 5 and the E to be suppressed. Values less than 1, on the other hand, will always have the sign given in the exponent so that 0.05, for instance, will be written as 0.5E−1.

7.14 Induced Error

7.14.1 When using this notation, which we must remember is *the normal way of manipulating real numbers*, there will be frequent induced errors from the decimal/bicimal fraction conversions, known as CONVERSION ERROR, which we looked at in Chapter 5.

7.14.2 A source of error which is typical of this system which we have not previously discussed is SUBTRACTIVE CANCELLATION ERROR resulting from the truncation of the mantissa when we need to find the difference between two nearly equal numbers in normalised floating-point arithmetic.

7.14.3 If we take, for example, 257.88 − 257.70 and this is normalised, by truncation due to register length, to 4 significant figures we shall have 0.2578E3 − 0.2577E3 giving a difference of 0.1 when restored to normal base 10 notation. The actual difference is 0.18 and E_r is therefore $100*(0.1/0.18) = 55.5\%$, an enormous error and one which we might well not expect. In order to reduce this to acceptable proportions we would be wise to use double precision, if available, or separate the calculation into $(257-257)+(0.88-0.7)$ once again making use of our knowledge of the laws to control the error in the machine.

7.14.4 Actually, modelling a routine like this is quite demanding. See if you can devise a sub-routine which would detect very small differences like this and pre-condition them for error diminution.

7.15 Conclusion

The nature of the machine makes many of our ordinary mathematical techniques inapplicable but, subject to the laws, we can use others which precisely suit the operation of the ALU.

New techniques need new understandings and experience if they are to become truly familiar to us. They also introduce new errors and limitations.

Exercises

1) Find the tens complement of:
   a) 563
   b) 999
   c) 101
   d) 23

2) Using tens complements calculate:
   a) 499–367
   b) 499–637
   c) 2495–792
   d) 1207–649.5

3) Code in BCD:
   a) 495
   b) 222
   c) 632
   d) 410

4) Calculate in BCD:
   a) 493+275
   b) 493–275
   c) 47+39+63
   d) 547+659+237

5) Convert to base 10 the partially calculated addition results in 'raw' BCD numbers:
   a) 0111 0011 0001 1001
   b) 1010 0001 0010
   c) 0100 1010 0110 0110
   d) 0111 0101 1011 1100

6) Express in 8-bit, twos complement form:
   a) 59
   b) 44
   c) 73
   d) 37

7) If the MSB has one sign bit and a seven bit exponent what is:
   a) i)   the largest,
       ii) the smallest number,
   which may be held in a 16-bit register?
   b) i)   the largest,
       ii) the smallest number,
   which may be held in a 32-bit register?

8) What is the practical importance of the difference in the answers to Question 7?

9) Show how a 16-bit register with one sign bit and a 7-bit exponent would hold a number reported as 0.4715E4.

8
Graphs, Functions and Inequalities

Objectives

After working through this chapter you should be able to:
- understand the need for graphical skills in computing;
- produce graphs to a high standard;
- generate graphs by spreadsheet;
- understand the significance of scales;
- relate functions to their appropriate types of graph;
- model continuous, linear, curved and scatter graphs;
- write appropriate computer program segments;
- use simultaneous equations and inequalities in graphs;
- assess error in graphs and graphical techniques.

8.1 A New Need

8.1.1 Until the early 1980s people working with computers had little need of any special knowledge of graphs, although these had long been a major modelling tool in mathematics. Our machines, in the main, delivered their processed output to devices which had little or no graphical capability. In specialist installations, large and costly plotters were used for engineering tasks of various kinds but the essential nature of the graph, *as a pictorial model of a mathematical situation or relationship(s)*, was largely irrelevant to data processing as we then knew it.

Within the following ten years changes in technology brought into common use high-resolution VDUs, dot matrix, laser and ink-jet printers, powerful software, fax machines and high-speed, high-fidelity data transmission systems to transform the world of information processing.

Visual models, as opposed to verbal or numeric outputs, have become common-place and a *major working tool for users at all levels*.

8.1.2 The old adage of journalism that ". . *a picture is worth ten thousand words*" is especially true of mathematics. Many people find difficulty in reading and interpreting large quantities of *numerical* information, even if it is arranged in the form of tables. They can see the implications of that information much more clearly if it is presented as a picture, which is what the word 'graph' means, quite literally. At a more sophisticated level we may construct graphs of relationships which allow us to derive considerably more information than we actually *put into* the graph whilst constructing it, or to solve quite subtle problems.

At this level the graph can provide us with an immensely powerful tool for analysis and prediction and the computer is well suited to generating, displaying, storing and transmitting such graphs.

8.2   Pictograms

8.2.1   Some of the world's earliest works of art, sophisticated paintings found in caves all over the world, may simply be graphs; records of the game found in the hunt for food or of the animals domesticated and used by the people who left us their pictures. Even today we still draw small images of what we want to record and when we group such images for the purpose of displaying or comparing situations we call the resulting graph a PICTOGRAM (see Figure 8.1).

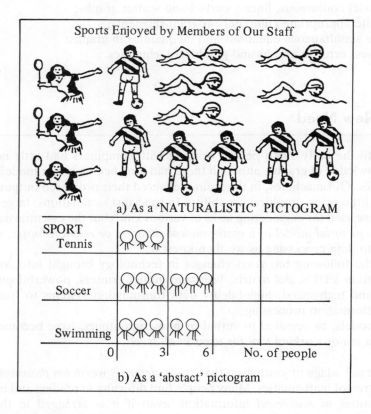

Figure 8.1   Two forms of pictogram

8.2.2 The images may be quite realistic or severely abstract but they confront us immediately with some of the basic problems of all graphs, indeed of all pictures. Although people often say, quite wrongly, "Figures cannot lie" we seldom add, ". . but pictures can and frequently do!"

8.3 Distortion

8.3.1 All pictures are themselves abstract models of what they represent and the person producing the picture makes decisions, consciously or unconsciously, about what to include, simplify, distort, omit or scale. Every artist does this, whatever medium is being used, and even in photographs the finest camera in the world still introduces its own misrepresentations into the images it creates.

8.3.2 If the person creating the image is a politician or an advertiser, the distortions may not necessarily be unintentional, as can be seen in Figures 8.2 and 8.3.

When we look at any graph we should ask ourselves whether the originator is simply modelling facts or is trying to persuade us to believe or buy something!

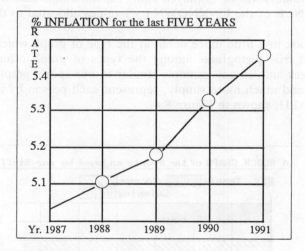

Figure 8.2 Truth or propaganda?

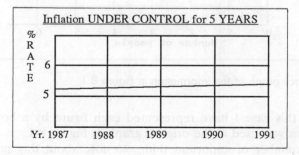

Figure 8.3 Government or opposition?

8.4   **Rules for Graphs**

8.4.1   In order to present graphs which are both useful and worthy of respect we must work within the discipline of mathematics. That is to say that we must follow certain rules, some of which are listed here; we will add others as they become necessary.

1) The type of graph chosen must be appropriate to the material being modelled.
2) We must give the graph a title which shows clearly what it is displaying.
3) We should add any sub-titles needed to clarify its purpose.
4) We must indicate clearly whether it is a one-, two- or three-dimensional representation.
5) We must define the scales of measurement used on each of the axes.
6) All scales must be clearly labelled and must be true to the range of values represented.
7) All calculated or input values must be clearly indicated and must be distinguished from *derived or inferred* values.
8) Unless otherwise stated TIME or the KNOWN values (often referred to as the '*x*' values) will be displayed along the *horizontal axis* with, for example, DISTANCE or the UNKNOWN ('*y*') values allocated to the *vertical axis*.

8.4.2   Let us look in a little more detail at the *type* of graph which is suitable. We shall rarely find true pictograms among the types of graph offered by computer software packages and, in any example similar to our sports graph, above, we can just as readily, and much more simply, represent each person by a simple 'box' on a BLOCK GRAPH, shown in Figure 8.4.

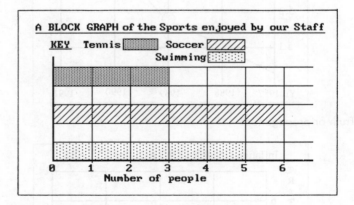

Figure 8.4   A block graph of the information in Figure 8.1

Note that in this case I have represented each figure by a box along the same horizontal axis as was used in the original graph in Figure 8.1.

There are a number of important things to note about this simple graph. It is, although drawn in two dimensions, a *one-dimensional* situation. Only the *number* of people in each category is significant and the categories must be indicated by either separate labels or by a 'key'. For ease of comparison the blocks all start from the same 'base-line' or DATUM, using the vertical axis for this purpose although the true ORIGIN of the graph, since it is one-dimensional, is the left-hand end of the horizontal axis. What we have done is to *extend the 0 datum, vertically, to serve as a common base*.

8.4.3   This is *not* at all the same as having a true vertical scale. The categories, tennis, soccer, swimming, which are *data labels*, rather than data, are also separate from each other, in mathematical language they are DISCRETE. We must, where this is the case, keep the piles of blocks separate from each other.

8.4.4   *Only where we are using a truly continuous scale may we place the blocks in contact with each other*.

Where this is the case the resulting graph is a HISTOGRAM, of which we shall see a good deal more later, but it is fundamentally different from a BLOCK GRAPH.

In order to make our graph look more familiar to us we may, instead of keeping to the original layout, make 'the floor' our datum line by plotting the graph with the *number of people* on the vertical axis. This is the more general arrangement for a block graph simply because gravity obliges us, in reality, to pile blocks in vertical, rather than horizontal, columns. Block graphs are essentially easy to draw and very clear. Have a look at Figure 8.5.

They are a little misleading since they suggest that it is the *area* of the block which is important. It is not. Only one dimension actually represents anything significant (and we should note that in our example this one true scale also gives *discrete* values, in this case integers, since we cannot have less than one whole person).

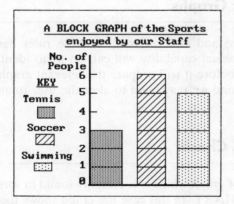

Figure 8.5   A block graph 'right way up'

8.4.5   At the lowest level of abstraction, our original figures, pursuing their sports, may be represented by a simple line *whose length indicates the single dimension, number*, with which we are concerned.

This type of graph also may be drawn with the lines either *horizontal*, usually called a BAR CHART (or graph), or *vertical*, when it is often referred to as a COLUMN GRAPH (see Figure 8.6).

8.4.6 These three types, the PICTOGRAM, BLOCK GRAPH and BAR CHART are simply *display* graphs; they model a fairly simple situation, are built up by plotting *all* the relevant data and are *interpretive only*.

We cannot infer from them anything which is not explicitly stated on the graph. Note, however, that our rules 2 to 6, above, must be obeyed in constructing them. This information is essential to giving the viewer a true picture of what is displayed.

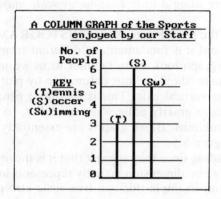

Figure 8.6 The same information simplified yet again

8.5 Spreadsheet Graphs

8.5.1 *We* must understand and get used to these rules because any spreadsheet system which has graphical capability will call on *us* to identify and input all these pieces of information before it will prepare the relevant graph. It may also not obey all the rules precisely and we may need to alert the user to any such peculiarities.

8.6 Proportional Charts

8.6.1 Another type of graph which is commonly found in spreadsheet systems is the PROPORTIONAL CHART. In this case the graph shows the whole of some entity, say *costs*, as in Figure 8.7, and subdivides this into the proportions contributed by various sub-divisions.

As always when working with proportions, it is most convenient to express the parts as *percentages*, as has been done here. This permits direct comparisons, year-by-year, or between similar companies, or divisions of a large organisation, in a way which is not nearly so easy if the various parts are expressed as common, or vulgar, fractions. If in one division of a company wages are 17 out of 23 $M, or 17/23 of total costs, whilst in another they are 19 out of 25 $M, (19/25) total, which has the proportionally higher wage-bill? If we convert the fractions to 74% and 76% respectively the comparison is easy.

COSTS(%)	1989	1990
Fuel	5	10
Materials	20	25
Overheads	20	15
Wages	40	35
Advertising	5	3
Salaries	10	12

Figure 8.7 A table of costs for two years

8.6.2 A simple way to graph such a table is to create a PROPORTIONAL BAR CHART where the data are displayed two-dimensionally (for ease of labelling, shading or colouring the different proportions) but only the length, (or height), is truly significant as the dimension which represents each of the individual percentages of the whole.

8.6.3 As we can see in Figure 8.8, the information is displayed clearly enough but it is not very easy to read from the graph the full *detail* of the changes which have taken place. We *can* see, for example, that the first three items have risen, in total, to 50% of overall costs and that proportionally less has been spent elsewhere. But it is difficult to assess from the graph exactly what has happened to the last three items of costs. A more detailed scale might help.

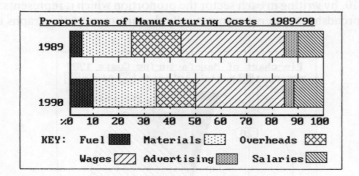

Figure 8.8 A proportional bar chart of our data

8.6.4 Another very popular form of the proportional graph is the PIE CHART. In this the different proportions are allocated correspondingly sized 'slices' of the whole pie. When our data are to be set into such a chart, we need to remember that the angle at the centre of a circle is 360° so that, if we are working with percentages, each 1% = 3.6° at the centre. Even this gives rise to some very awkward angles if we are trying to measure them by protractor, but vulgar fraction proportions can be very awkward indeed. How many degrees at the centre would 17/23 represent?

8.6.5   The result is that measurements on pie charts are always, in practice, likely to be approximations.

8.6.6   In Figure 8.9, we see that, once again, it is difficult to read the exact significance of the different sizes of sectors and this is unavoidable even if we happen to have a protractor handy when we are attempting to interpret the graph.

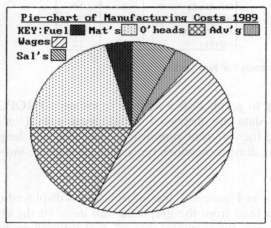

Figure 8.9   Our data in a pie-chart

8.6.7   Such graphs give a reasonable impression of the data from which they were constructed but are really suitable for display purposes rather than for any subtle interpretation. We can improve matters somewhat by *annotating* the graph, as shown in Figure 8.10, by writing in each sector the proportion which it represents. In practice, almost all spreadsheet systems automatically do this to make the graphs more legible to the user.

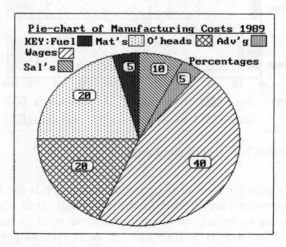

Figure 8.10   An annotated pie-chart

8.6.8 A final form of this type of graph is what is known as an EXPLODED PIE-CHART where, in order to emphasise the significance of one or more elements of the graph we show the 'slice' or slices concerned as partly withdrawn from the whole pie. This can be seen in Figure 8.11.

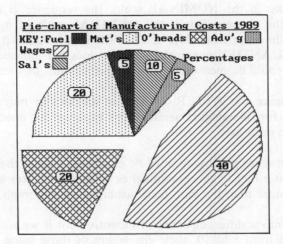

Figure 8.11 An 'exploded' pie-chart

8.6.9 Once again, many spreadsheet systems offer the option of an exploded pie-chart. It is worth nothing that some writers, clearly believing that there can be nothing simple about mathematics, insist on calling these 'PI' charts, using the verbal form of the Greek symbol π, presumably because this has something to do with circles! This is a strange form of snobbery and has nothing at all to do with the nature of these graphs.

8.7 Scales

8.7.1 Before passing to other types of graph we need to look at the matter of scales, item (5) on our list. In order to do this we must also examine more closely the nature of measurement, which may be on a NOMINAL, ORDINAL, INTERVAL or RATIO scale.

8.7.2 For the first of these we are using numbers in a rather deceptive way, simply as labels under which to *classify* items. We may say, for example, that in a survey of sports, all tennis players should be put into class 1, all soccer players into class 2 and all swimmers into class 3 and we could then number the blocks or bars of a graph accordingly. We call such numbers NOMINAL measures and we must be extremely careful always to treat them as just the *names* of classes and NOT as true numbers.

8.7.3 Such measures are often useful, especially in some kinds of statistical tests, but must always be treated as NON-PARAMETRIC.

8.7.4 If we take a different approach and ask 100 people to classify different sports *in order of preference* and the verdict emerges, tennis 1, soccer 2, swimming 3, it is clear that, unlike in the NOMINAL scale, the sequence *is* important since it has ORDERED the classes. This is the ORDINAL scale of measurement which we frequently use in order to RANK or ORDER items. Again it is non-parametric and it does not QUANTIFY our preferences. It does not indicate by *how much*, if such a thing even *could* be measured, people prefer tennis to soccer.

8.7.5 When we make a SCATTER GRAPH in statistics we may plot, for example, students' rankings for one subject against their rankings for another subject, using an *ordinal scale* on each axis.

8.7.6 When we prepare graphs we must be extremely careful that our presentation does not imply that either of these nominal or ordinal measures alone may be used to infer anything more than the bare facts displayed on the graph.

8.7.7 We move into a different order of measurement if we ask the same people to rate, on a scale from 1 to 10, their enjoyment of those same three sports. It is then implied that each INTERVAL on the scale has the same magnitude and that someone rating tennis as 10 and soccer as 2 would consider the former to be 5 times as pleasurable as the latter.

8.7.8 Rather similarly we would be entitled to assume that someone with an Intelligence Quotient, (IQ), of 110 was as far *above* the 'norm' of 100 as is someone with an IQ of 90 *below* that same norm. These are two applications of an INTERVAL SCALE of measurement, where there is exactly the same *interval* or range of value between any two marked points.

8.7.9 The important difference is that the first example is SUBJECTIVE, or is a matter of feeling or opinion, whilst the latter, since it purports to be based on actual *measurement*, is OBJECTIVE. We must always recognise AND DECLARE ON OUR GRAPHS, by means of a subtitle perhaps, if the information being displayed is *subjective*.

8.7.10 One of the special features of many interval scales, the IQ scale for instance, is that they take a mid-point or 'norm' and extend on either side of that. There is no 'zero IQ', only more or less than the norm of 100. We find many such scales in statistical tests and graphs and shall see more of them in that context later.

8.7.11 The most refined form, or the HIGHEST ORDER of measurement is that on a TRUE RATIO SCALE. For all of these we deal in OBJECTIVE VALUES on a scale which has a TRUE ZERO. Virtually all measurements made with instruments are of this nature and many of them originate at zero and continue to infinity. They may also extend to the negative as well as the positive numbers.

8.7.12   The final point about all such scales is that they may be DISCRETE or CONTINUOUS. When we are counting the number of people at a football match, for example, we are using a RATIO scale, since there can be a true zero, but each person is a separate entity and there are no fractions, we use a DISCRETE RATIO SCALE. If, on the other hand, we are plotting a graph of a sick patient's temperature the mercury in the thermometer can vary infinitely between say 32° and 33°C, and the variations are on a CONTINUOUS RATIO SCALE.

This is one of the most important and least understood aspects of measurement and consequently of both the presentation and interpretation of graphs. *Many spreadsheet and other graph-drawing packages are quite unable to distinguish between discrete and continuous data* and *we* may need to use sub-titles to draw attention to the true nature of our graphs.

8.7.13   Let us look at a specimen graph in order to clarify the essential differences between a discrete and a continuous scale. If we take the sales chart in Figure 8.12 we see that, represented as a bar-graph, we are looking at a *discrete* scale on the horizontal axis.

The vertical axis is a true continuous ratio scale since any amount may be represented.

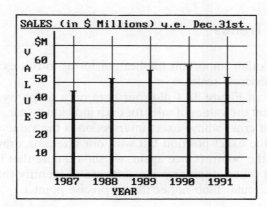

Figure 8.12   Sales as bar-chart

8.7.14   When we display the same data as a block graph, however, since there seems to be a continuous time-scale from 1987 to 1991, it might seem that we could place our blocks side-by-side as in Figure 8.13.

8.7.15   This is not strictly true since the scale is discrete and there is actually a gap of 364 days between Dec 31 1987 and Dec 31 1988 – and likewise throughout the horizontal scale. The totals for sales are not recorded day-by-day in our data (although they undoubtedly will have been within the accounts department of the company concerned). All that we have is the TOTAL OF SALES, accumulated throughout the year, without any knowledge of their details. This also means, as we saw earlier, that it is only the *heights* of the blocks which is significant.

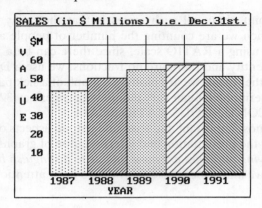

Figure 8.13 The same data as block graph

8.7.16 Even if we do close the gaps between the gaps, as in Figure 8.13, this *is not a HISTOGRAM*, it is still only a block graph.

8.8 Points and Trends

8.8.1 There is one other important implication for us in this example, yet another consequence of the discrete scale.

We could well, as in Figure 8.14 plot our data on a graph by marking the exact points, which represent the values of sales for each individual year. We usually identify each point by a small cross whose exact intersection is the value concerned, or by a dot, again marking the exact position but with our attention drawn to it by placing a small circle round the point. (Once again, we should note that it is not uncommon for spreadsheets and other graph-drawing software to identify the point by drawing a small square or a circular blob rather than a precise point.)

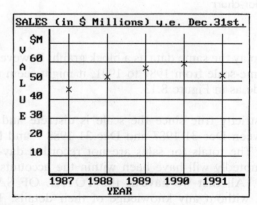

Figure 8.14 A 'point' graph

8.8.2   Having plotted our points we must not, however, then 'connect' them with a *continuous* line.

8.8.3   Such a line defines the *relationship* as CONTINUOUS, as we shall see more clearly when we look at the graphs of functions presently.

There is no evidence whatever that sales rose *either steadily or continuously* between Dec 31 1988 and Dec 31 1989. They will in all probability have fluctuated, perhaps considerably.

8.8.4   All that we may say, and we must use clearly *dis-continuous* lines on our graph in order to say it as in Figure 8.15, is that *there was a general trend for sales to rise* between those two dates, taking the year as a whole. This is another of the strict rules which is often broken by spreadsheet systems which tend to treat *all* line graphs as continuous.

8.8.5   Such 'TREND-LINE' graphs are actually of great value in management and we may, at a more sophisticated level, analyse and continue the trend beyond the known data.

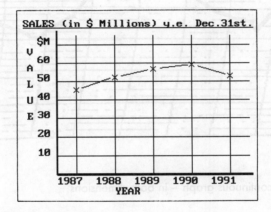

Figure 8.15   'Trend' lines for discrete data

8.9   Continuous Functions

8.9.1   We may only use a continuous line on a graph when we have clear evidence of the continuity of the function but in that case we *must* use a continuous line, which truly follows the function. Providing these rules are obeyed we may then INTERPOLATE (read values *between* plotted points) and EXTRAPOLATE, or read values *beyond* the plotted points.

8.9.2   The example shown in Figure 8.16 is of truly continuous ratio scale graph where the pen which draws the line on the graph, indicated by the arrow, is continually moving in both axes. Along the horizontal axis the paper, wound on to a drum, is moved by a clock mechanism in real time. Movement in the vertical axis is continuous because the pen is mounted on a swinging arm which rises and falls in real time as the atmospheric pressure changes.

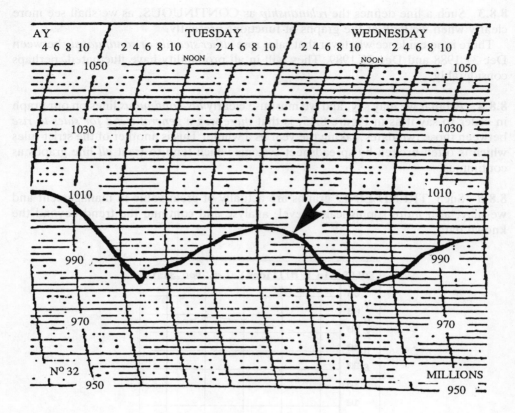

Figure 8.16   A truly continuous graph – in both dimensions

8.9.3   Such continuous graphs are relatively common in computer controlled industrial processes where the machine is continuously reading two or more parameters and recording them by plotter or by a continuous line of overlapping dots.

In this case the variation in height along the '*y*' axis is a naturally occurring phenomenon and cannot be simply predicted. This is no simple two-variable relationship but the very complex association of many variables in the world weather system.

8.10   Graphs of Functions

8.10.1   Happily, at this level we are not called upon to deal with such complicated models but there *are* three classes of simpler two-variable functions on continuous ratio scales which we can and should be able to model.

8.10.2   They are:

1) $y = mx + c$ (first order)
2) $y = m(x^2) + c$ (second order)
3) $y = m(x^3) + c$ (third order)

and variations on, or combinations between any of the three.

8.10.3   The three types of function, the LINEAR, QUADRATIC and CUBIC respectively have much in common since each of the values of x may be multiplied by the value 'm', which may be positive, negative or 1. Each may also have added a constant value, 'c', which may again be positive or negative but may also be zero. You will, I am sure, remember that in the laws of IDENTITY 1 and 0 are the identity elements for multiplication and addition respectively; they leave the value of the 'x' unchanged.

8.10.4   In:

$y = 2x$ the $m=2$ and the $c=0$
$y = x - 5$ the $m=1$ and the $c=-5$
$y = 2x^2-7$ the $m=2$ and the $c=-7$

and so on. The 'm' indicates the steepness of the curve on the graph or the *ratio* or gradient (the rate of change) of the slope, often expressed as a RATIO. For example 4:1 if the 'y' value increases at *four times the rate* of the change in 'x' value. In that case 'm' = 4.

8.10.5   The 'c' indicates the height, above or below 0, at which the curve 'cuts' the 'y' axis; this is known as the *intercept or displacement*.

8.10.6   We may also have various combinations of elements as in

$$y = 3x^2 + 2x - 7$$

which do not add any great complication to the model but may just look a little more forbidding.

8.10.7   One such function which looks a little complicated but produces a slightly surprising result when plotted is of the type $x^2 + y^2 = 4$. Try this one later and see what happens.

8.10.8   The three types of function show characteristic curves on the graph fields (which can be seen in Figure 8.17) and their actual steepness and positioning will vary according to the actual parameters of 'm' and 'c'.

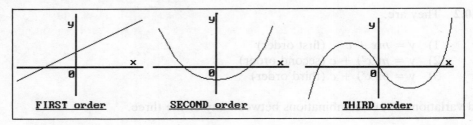

Figure 8.17 The three characteristic graph curves

8.11 Some Further Rules

8.11.1 We can, once again, draw up some general rules for the presentation of the graphs.

1) Unless the situation expressly excludes them, for example where we cannot have negative entities in the universe of discourse concerned, the scales on both axes should embody at least some part of the negative values in their ranges.

2) Unless the set of values on the scale concerned is *finite*, the scales should be shown, in each axis, to extend to infinity and at least be continued right to the edge of the paper.

3) The scale actually chosen for the graph should be the largest which will cover, on the size of sheet available, the range of values stipulated so that we may have the benefit of the largest plot which can be set out on the available sheet of paper. (If, for example, we are told that the '*x*' values should extend from –3 to +3 the whole width of the '*x*' axis should be used for, say, 7 units so that we may have the smallest induced error in our graph.)

4) The marking of points and the drawing of lines should be as fine as possible to minimise induced error.

5) For a linear function three points should be plotted as a check on the arithmetical accuracy of the minimum two points. If the three points do *not* fall accurately on a straight line one of them has been wrongly calculated.

6) For all second and third order functions the line of the graph, no matter how few points have been plotted, is a *continuous smooth curve*. The points must on no account be joined by straight lines. The more points we plot the easier it is to smooth the curve, especially with spreadsheet systems where lines between plotted points are almost always straight!

7) We should label the curve on the graph with the function which it represents. This becomes especially important when we have two or more functions plotted simultaneously on the same graph field.

8) Only if all these conditions have been complied with can we use the graph for reliable inter- and extra-polation.

8.11.2   For the actual plotting of points it has long been customary, and still is for *manually* drawn graphs, to set out the calculation of the points for plotting a function $y = 2x^2 + 2x - 3$ such as is shown in Table 8.1.

Table 8.1

If	$x =$	-3	-2	-1	0	1	2	3
then	$2x^2 =$	18	8	2	0	2	8	18
and	$2x =$	-6	-4	-2	0	2	4	6
and	$-3 =$	-3	-3	-3	-3	-3	-3	-3
hence	$y =$	9	1	-3	-3	1	9	21

8.11.3   This method is still very reliable and, as with the $y = mx + c$ graph, if any one of the points does *not* fit to our carefully drawn *smooth curve* it is likely to be the arithmetic which is at fault.

8.12   Graphs by Computer

8.12.1   Such an elaborate table is not needed for the computer nor is it desirable. Remember that the great advantage of the machine is that it will, without becoming bored or impatient, do large numbers of repeated calculations in very little time. For us this has many advantages since we can calculate many points and derive a much smoother curve from a machine which resolutely tries to draw *straight lines* wherever possible.

For the example above, where we wish to plot our curve for values of $-3 \leqslant x \leqslant + 3$, we could as easily take intermediate values at intervals of .01 as of 1. The six hundred separate calculations would take little time in the machine.

```
X, Y REAL
DIM X(601); Y(600); j=1; X₁=-3;
DO UNTIL Xⱼ=3.01
        Yⱼ ← 2*(Xⱼ**2) + (2*Xⱼ) -3
        j ← j+1
        Xⱼ ← Xⱼ₋₁ + 0.01
ENDDO
DO PLOT
END
```

is just one sketch of a pseudo-code routine which would calculate all the values in readiness for calling a 'PLOT' routine to print the curve either on screen or on paper.

8.13   The Area Under a Curve

8.13.1   One of the ancient problems with such 'curved' functions has been to find the area under the curve between any given pair of values and the rather elaborate formulae of the past gave us, very laboriously, approximations of that area. By a small modification of the routine, suggested above, for plotting the points we can dispense with such manual labours today. The digital computer is the machine which Leibnitz and Newton tried to emulate by the tedious plodding of the calculus!

8.13.2   Between any two successive values of '*x*' the area of the narrow vertical strip beneath the line of the graph is effectively a rectangle whose width is the constant 'step' size (ie, the difference between the two '*x*' values) and whose height is the average of the two corresponding calculated '*y*' values.

We can append this extra calculation, of the area of each strip, to the existing routine and on each pass of the loop add each new area to a variable, say 'A', as a CUMULATIVE total. The computer, with its highspeed iterative methods has in such ways rendered the calculus virtually obsolete for most practical purposes.

8.14   Interpolation and Extrapolation

8.14.1   From any of our plotted functions we can derive the values represented by any other point on the line. For any given value of '*x*' we raise a vertical line to the point where it intercepts the curve of the function and by projecting a horizontal line to the '*y*' axis we may read off the corresponding '*y*' value.

8.14.2   By the inverse process we can find, from any known '*y*' value, the '*x*' value from which it is derived. Since this is a quite common requirement it is often well worth adding such an 'enquiry' routine to the curve generating routine of the function. Whether we wish to show such interpolations on the VDU screen will influence the way in which we set up such routines. You should experiment with such procedures.

8.15   Simultaneous Functions

8.15.1   The next important development for us is when, for a given situation, two or more functions are simultaneously true.

8.15.2   If, for example, we have two equations of the $y = mx + c$, first order type in which the '*m*' is different in each case, each will have a straight line on the graph and, if the functions are infinite those lines must intersect *somewhere* on our graph field so long as we have chosen the correct scale to reveal that intersection. You can see this in Figure 8.18.

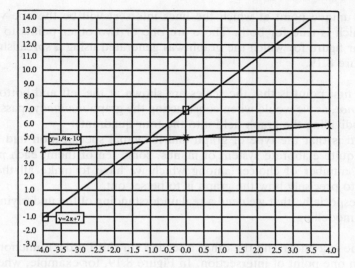

Figure 8.18 The intersection of $Y = 2X + 7$ and $Y = \frac{1}{4}x - 10$

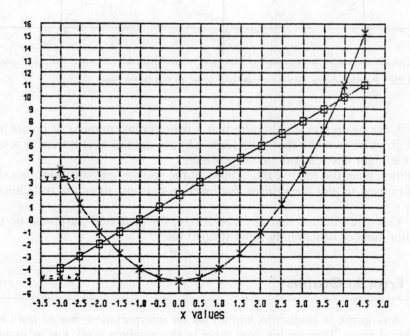

Figure 8.19 The graph of $Y = 2X + 2$ *and* $Y = X^2 - 5$

8.15.3   The unique point at which the lines intersect will be the one value of the functions which is true for both. There are one or two other points to note about *this* particular figure for which the graph was generated using a spreadsheet system. See also Figure 8.19.

8.15.4   The first point is that the scales are shown at the left and bottom edges of the plot and *not*, as we would when constructing the graph *manually*, passing through zero and subdividing the graph field into the four quadrants.

The second is that the type of graph and its actual specification had to be input through the quite elaborate system of menus, some ten of them, each presenting a considerable number of choices among which *we* have to make all the necessary decisions as to precisely how the graph is to be set out.

It is here, especially, that we need a real understanding of the underlying principles of graphical modelling.

8.15.5   If one is a linear and the other a second (or third order) function there may be more than one point of intersection. In Figure 8.19, for example, where we have the plot of *both* $y = 2x+2$ *and* $y=x^2-5$, we see that there are two points of intersection and in that case *each of those points* will be common to the two functions and the common values of '*x*', '*y*' or both may be read off the graph.

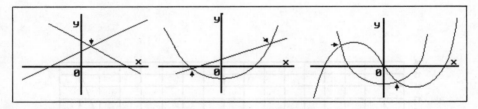

Figure 8.20   Points where two functions are 'true' at the same time

8.15.6   Once again you should note that a considerable number of points has been plotted, in order to make the curve (drawn by the spreadsheet system as a sequence of short straight lines) as smooth as possible.

It follows, from the relative inflexibility of the spreadsheet software, that the same points are used to plot the *straight line* function but only three are truly necessary.

8.15.7   Can you also see that any curve may, for analytical purposes, be taken as an infinite number of infinitely short straight lines?

8.16   Error in Graphs

8.16.1   Any graph is subject to limitations on interpretation due to both inherent and induced error. There is inherent error in the numbers used, just as in any other application, but there is also error induced by our technique.

8.16.2 In all cases we shall find that large-scale plotting is needed to minimise induced error since both the thickness of the lines and the angle at which the lines cross may induce prohibitively large error if the graph is on too small a scale. Figure 8.21 shows the very small area of uncertainty when very fine lines cross at right angles, or close to it. It is so small that in the diagram it has been 'magnified' to three times its true area to make it clear. On the right of the same diagram we see the very large area of uncertainty, the error bound, which is induced if badly drawn, *thick*, lines cross at a 'shallow' angle.

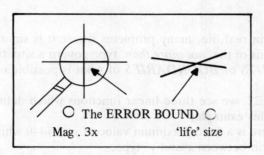

Figure 8.21   Error induced

8.16.3 The closer the 'crossing' of the lines is to a right angle the smaller the error; the larger the scale the smaller the error induced by the thickness of the lines themselves. Even the lines created by the printer of the graph-paper induce error and it pays to use the very best quality of graph paper for the really important graphs.

8.16.4 In practical terms, in computer-generated graphs, we need to be able to take account of inherent dot size in the printer, vertical movement of the paper when printing and the number of dots per inch which can be printed, both vertically and horizontally.

These things come under the heading of 'resolution' when we talk of print quality and accuracy and it is well worth becoming familiar with the capability and the limitations of the printers and the screens we are using.

8.17   Inequalities

8.17.1 The last aspect of graphs, for the present, is the displaying and solution of problems concerning simultaneous *in*equalities.

8.17.2 We saw, much earlier in the book, that any line on a graph embodies the set of points which is true of the equivalence function. So the graph of $y = 2x + 2$ will embody, in the line we draw, the infinite set of points for which that equivalence holds good.

8.17.3   Every such continuous ratio scale function has its own infinite set of such points. By defining such a set we also, and inevitably, define the *inequalities* which are also created. They are:

a) $y > 2x + 2$; b) $y < 2x + 2$;
c) $y \leqslant 2x + 2$ and d) $y \geqslant 2x + 2$;

of which the last two *include* the line of the function itself whilst the first two are the regions, or domains, *above* (but *excluding* the line) and *below* (but *excluding*) the line respectively.

8.17.4   There are, in real life, many problems where it is simultaneously true that two or more functions of the '*not more than*' type govern a situation, often by setting or defining the *LIMITS* or *BOUNDARIES* of what is possible or feasible.

8.17.5   In Figure 8.22, we see three linear functions which define limits which *may not be exceeded* in this example.

For function 1 there is a fixed maximum value of $y \leqslant 1440$ whilst each of the other is of the 'inverse ratio between x and y' type.

The precise nature of the functions need not concern us in this example of such a model, but what *does* concern us is that the area under and including each of the three lines represents the '*not more than*' region in each case so that *only the region which is below and including (sections of) all three lines together satisfies all three constraints simultaneously.*

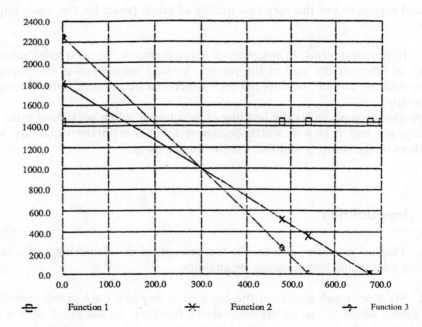

Figure 8.22   The boundary of a *feasible* region

8.17.6   The three function lines together define a polygon within which any possible feasible solution must lie. We call this, in practice, the FEASIBILITY REGION. If, using ordered pairs for the points which define the polygon, we start at (0,0) and follow the sides of the polygon clock wise the subsequent points are approximately: (0,1440); (135,1440); (300,1000); (540,0) and back to (0,0).

8.17.7   Many management problems can be modelled in this way on a spreadsheet system since such problems are often constrained by several 'not more than' functions, manufacturing capacity for instance. The great advantage of using a computer spreadsheet graphical model is that we can easily change the various functions on a 'what if' basis and quickly see the effects. At the very least we can print out the graphical models so that managers can *see* the area of manoeuvre which is open to them.

8.18   Conclusion

Graphs are a vivid means of modelling real situations, give us a powerful tool for analysing situations and allow us to interpolate or extrapolate solutions to many different kinds of problem. They fit well with the computer display on the VDU or by high-resolution print-out.

Exercises

1) Draw appropriate graphs to display:

a) A child's height in cm:

Age	2	3	4	5	6	7	8	9	10	11	12	13	14
cm	87	98	107	112	117	123	127	131	135	140	145	149	154

b) A company's sales, over a 16-week period:

WEEK	10	11	12	13	14	15	16	17	18	19	20	21	22	23	24	25
Sales 000's	3.2	2.3	4	4.3	3	4.8	4.4	4.4	3.6	5.2	3	4	2.9	4.6	4.7	5

2) a) Comment on the graphs in Question 1).
   b) Would it be proper or valuable to interpolate on either graph?

3) a) Over a three-month period the earnings in a garage were $37500 from petrol sales, $21100 from repairs, $1200 from the car hire service and $19900 from car sales. Present these data as a pie chart.
   b) In the following three months petrol sales were $42900 out of a total of $87,500. Represent this as an exploded pie chart.
   c) By how much, as a proportion of the whole earnings, did petrol sales increase between the first and the second three-month periods?

4) What are the essential differences between a column (or block) graph and a HISTOGRAM?

5) a) From the data in question 1 develop the two moving average graphs.
   b) Insert a trend line into each graph if appropriate.

6) For $-4 \leqslant x \leqslant 4$ calculate and plot on a graph the function
$$y = 2x^2 + 2x + 4$$

7) a) On a single graph plot the functions $y = x^3+4$ and $y = 27-2x$
   b) At what values of 'x' and 'y' do the functions intersect?
   c) For $0 \leqslant x \leqslant 3$ and $y \geqslant 0$ show, by shading, the feasibility region for $y \leqslant x^3+4$ and $y \leqslant 27-2x$.

8) A small manufacturer of steel boxes needs to produce no more than he can sell, or than he can store until the product is actually sold. His box-making machine can be set up to produce *either* large or small boxes at any one time but not both.

   The available storage space totals 4 cubic metres; large boxes occupy $1/100$ m^3 whereas small boxes take up 0.008m^3.

   He can sell, at most, 280 large, or 400 small boxes per day and in each 8-hour day the machine can produce either large boxes at 40 per hour, or small boxes at 80 per hour.

   a) Construct a model of each of the restricting or constraining functions, for example,

   $$4 \geqslant 0.01L + 0.0085S$$

   b) Calculate the three inequalities for an appropriate range of quantities of large and small boxes.
   c) Plot the three functions on a graph and outline the region which identifies the possible allocations of production between large and small boxes.

9
Matrices and Graphics

Objectives

After working through this chapter you should be able to:
- relate shapes and their movements in space to the computer;
- use elementary vectors;
- understand the motion and transforms of 2D shapes;
- model shapes and operations on them by matrix arithmetic;
- perform a range of elementary 2D spatial operations.

9.1   The New Graphics

9.1.1   To anyone accustomed only to traditional data processing, a first encounter with DTP, CAD or CAM, much less computerised navigation systems or automatic pilots, may come as a great shock. How can the computer manipulate *spatial* information of such complexity and at such speed, or cause metal to be cut with such accuracy?

9.1.2   The concept of the graph, the graphical capabilities of the VDU screen and the finely controllable printer or plotter, combine to open up some of the most widely useful ranges of computer applications.

We exploit the ability to emulate and to manipulate entities in two or three-dimensional space and to model real objects and relationships in spatial terms. Ultimately these same basic ideas extend to controlling on the one hand, the machine tools which cut or weld metal, on the other, the fly-by-wire aircraft.

Essentially, it is the computer programmer who plans, through the program(s), and thereafter controls, *movement in three-dimensional space and time*, whether it be of the pencil, the aircraft, or the cutting edge of a machine tool.

9.1.3   As we have already seen the essential *spatial* modelling tool is the CARTESIAN SPACE, a square grid on which the precise location of any point is determined by identifying the quadrant, by the 'plus' and 'minus' signs of an ORDERED PAIR, of which the numbers then give the actual position within the quadrant. (Note that the four quadrants give four possible sequences of $\pm x, \pm y$.)

The quadrant numbered 1 in Figure 9.1 is sometimes referred to as the '+,+' quadrant whereas that numbered 3 is that '−,−' quadrant and so on.

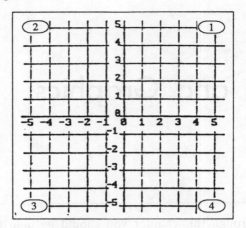

Figure 9.1 The four-quadrant Cartesian space

9.1.4 In the example in Figure 9.2 the point A, the intersection at the centre of the circle (which is used as an *indicator* only) is defined, without any ambiguity, by an ordered pair of the general type $(\pm x, \pm y)$, with the 'x' value, or the horizontal offset from zero, always given first and followed by the 'y' component, the vertical offset from zero. Point 'A' is (7,3), point 'B' is (−6,2), point 'C' is (−5,−5) and 'D' is (3,−4).

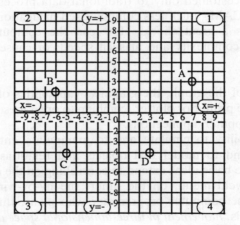

Figure 9.2 Points on the grid

9.2 Vectors

9.2.1 In Figure 9.3 we see that the 'journey' from A to B may be indicated by a straight line, which, however, needs an arrow to make clear that it is from A to B and *not* from B to A. Such a line indicates a CHANGE OF, or DIFFERENCE IN,

POSITION and, since a DIFFERENCE is the result of a *subtraction*, may be found by subtracting the first ordered pair, the starting position, from the second such pair, that of the finishing position: $(-6,2) - (7,3) = (-13,-1)$.

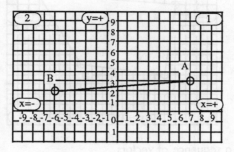

Figure 9.3   The 'journey' from A to B – a vector

9.2.2   The DIFFERENCE BETWEEN ANY TWO POINTS (or any quantum of which we know both magnitude and direction) may be indicated by such a VECTOR which, in the present case, (and as we saw in Chapter 7), also identifies a continuous linear function of SLOPE or RATIO $-1:-13$, which has, in the *opposite direction*, B→A, the *inverse slope* of 1 : 13. This tells us that for a change of 13 on the 'x' axis the change on the 'y' axis is 1. We saw in the previous chapter just how easy it is to program the computer to calculate intermediate points of such functions, no matter how small we wish to make the step size, and to draw the resulting lines.

9.3   Vectors and Shapes

9.3.1   If we link *all* the points in sequence, A to B, B to C, C to D, D to A, with a succession of similar vectors, as is done in some elementary programming languages such as LOGO, we find that we have described a *shape*, in this case an *irregular quadrilateral*. See Figure 9.4.

In narrative form we could say, "If we start at the point (7,3), move 13 left and down 1; 1 right and down 6; right 8 and up 1; right 4 and up 6; we describe the shape ABCD." Notice how '+' equates with 'move to the right' and '−' is 'left' along the 'x' axis but 'up' and 'down' respectively on the 'y' axis.

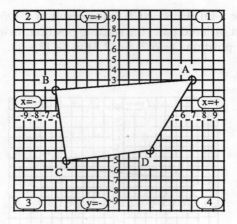

Figure 9.4   A shape as a sequence of vectors

9.3.2   These designations of movement or change of position are of great importance and we need to practise using them in a variety of situations.

9.3.3   This 'shape' also describes a *journey* from A to B to C to D and back to the start at A with the four 'legs' of the trip as the vectors:

$$\overline{AB} , \quad \overline{BC} , \quad \overline{CD} , \quad \overline{DA} .$$

9.4   Notation

9.4.1   In this notation we see that the *sequence* of letters shows the *direction* whilst the line *above* the letters indicates a VECTOR, whose direction and length are both known (rather than a LINE SEGMENT, where length only is known, or a RAY, of which we know the direction only). Both the concept of the vector and the notation are of great importance in many applications.

9.5   Combining Vectors

9.5.1   We may also COMBINE VECTORS. In our example the journey from A to B and then to C is equivalent, in terms of overall progress, to going *straight* from A to C. In vector terms $\overline{AB}+\overline{BC}=\overline{AC}$. Let us see if the arithmetic works out in ordered pairs.

A to B = $(-13,-1)$ *and* (or +) B to C = $(1,-6)$ and adding these together we have $(-12,-7)$. A direct path from A to C shows the *difference* of $(-5,-4) - (7,3)$ which is, perhaps unsurprisingly, $(-12,-7)$. The result of combining two or more vectors like this we call the RESULTANT, again a very useful little tool.

9.5.2 You may well have worked such practical examples as, for instance, finding the actual path of a boat when a rower can achieve, in still water, a speed of six kilometers per hour but the river flows at eight kph. In the graphical model of the rower attempting to travel straight across the river, in Figure 9.5, the river is envisaged as flowing from the bottom of the graph to the top with the 'near' bank along the '*y*' axis.

The rower, starting at (0,0), could if there were no current, row straight across from A to B, represented by the vector AB with an arrow indicating direction. The vector is shown here as a thick line for clarity of illustration only since a vector normally *has no thickness*.

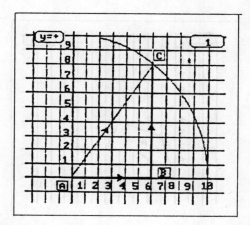

Figure 9.5 Modelling the river

9.5.3 The current, flowing from B to C, is also shown as a (thick line) vector. The result of combining *both* the speeds *and* the directions is the vector AC which shows that the boat would actually progress downstream at the angle CAB, rather than straight across the river.

By striking the arc of a circle from C to the horizontal scale we can also determine that the resultant *speed* would be 10 kph. Where would the rower have to 'aim' his boat and how fast would he have to row in order to make his boat travel straight across the river at 6 kph?

What would happen if the rower were to row faster, or slower? Try some variations on the original vectors and plot the resultants; as always your work should include trying several such examples until you are confident that you understand.

9.6 The Numerical Model

9.6.1 The final modelling tool in such applications is the MATRIX containing the NUMERICAL MODEL, or set of ordered pairs, which may then be manipulated by well tried and proven mathematical techniques within existing laws.

9.6.2 To revert to a shape model (and we will confine ourselves to two dimensions only) Figure 9.6 is a simple shape which is, quite deliberately, non-symmetrical.

It is five-sided and so our matrix representation of it will have five ordered pairs.

$$A = \begin{bmatrix} 0 & 3 & 5 & 4 & 1 \\ 0 & 2 & 2 & 4 & 3 \end{bmatrix}$$

9.6.3 We have, in this way, modelled our irregular pentagon in a form which is a very economical data structure and which we can operate upon by conventional matrix arithmetic.

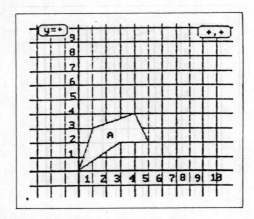

Figure 9.6 The five-sided shape for exploration

9.6.4 But what kinds of operation may we need to perform?

In the real world objects are not static but dynamic. They move, grow, shrink and change their appearance in ways which traditional geometry does not envisage.

9.7 Transformations

9.7.1 It is not surprising that there is a GEOMETRY of TRANSFORMATIONS, part of MOTION GEOMETRY, in which we examine and operate upon shapes which are *dynamic*. The computer, more than any other of man's inventions, is a wonderful tool for modelling the dynamic world.

9.7.2 Let us start with some very simple changes.

9.7.3 In Figure 9.7 we see an ENLARGEMENT: our original pentagon, 'A' in this diagram, is enlarged to become A′ which is clearly the same shape although of different size. Mathematically we identify this as SIMILAR, the shape is identical but the size is different.

Our original matrix:

$$A = \begin{bmatrix} 0 & 3 & 5 & 4 & 1 \\ 0 & 2 & 2 & 4 & 3 \end{bmatrix} \text{ has become } A' = \begin{bmatrix} 0 & 6 & 10 & 8 & 2 \\ 0 & 4 & 4 & 8 & 6 \end{bmatrix}$$

9.7.4 What is the mapping function which has converted 'A' to A'? It does not take a great deal of investigation to reveal that all the pairs which make up 'A' have been multiplied by 2 to generate A'.

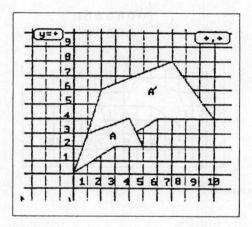

Figure 9.7 The shape enlarged

9.7.5 Such an enlargement, in matrix mathematics, is known as a 'scaling' and the multiplier is described as a SCALAR. We write the operation as:

$$2 * \begin{bmatrix} 0 & 3 & 5 & 4 & 1 \\ 0 & 2 & 2 & 4 & 3 \end{bmatrix} = \begin{bmatrix} 0 & 6 & 10 & 8 & 1 \\ 0 & 4 & 4 & 8 & 3 \end{bmatrix}$$

What happens if our scalar is ½? Try various *other* scalings and see what happens. (You may need to change the scale on which you construct your graph in some cases in order to see clearly what is taking place.)

9.7.6 In general terms we talk of all variations on an original shape, or its matrix, as TRANSFORMS and the next such 'family' to consider are the TRANSLATIONS. In a *translation* from one language to another we 'transfer' a piece of spoken or written text from the one language to the other *without changing its sense*.

9.7.7 A mathematical translation is very much the same; the shape is transferred to a new position without changing its appearance in any way.

9.7.8 Let us look first at Figure 9.8, a simple *lateral translation* in which our shape is merely 'slid' sideways.

Our original matrix,

$$A = \begin{bmatrix} 0 & 3 & 5 & 4 & 1 \\ 0 & 2 & 2 & 4 & 3 \end{bmatrix}$$

becomes:

$$A' = \begin{bmatrix} 6 & 9 & 11 & 10 & 7 \\ 0 & 2 & 2 & 4 & 3 \end{bmatrix}$$

and, in comparing the two to see what *difference there is* between them, by *subtracting* the first matrix from the second, we find the matrix (with Δ as the symbol for *difference*), in which we have five repetitions of the vector (6,0).

$$\Delta = \begin{bmatrix} 6 & 6 & 6 & 6 & 6 & 6 \\ 0 & 0 & 0 & 0 & 0 & 0 \end{bmatrix}$$

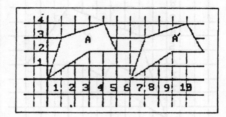

Figure 9.8   A lateral translation

9.7.9   There is, however, a LAW of IDENTITY for vectors and matrices which says that, "If two or more vectors or matrices are identical *they are the same vector, or matrix*." For an illustration of this look at the transform in Figure 9.9.

For the movement of each corner of the shape to its new position, check the vector by subtracting each starting point from its relative finishing point.

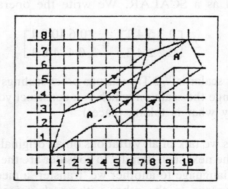

Figure 9.9   The common vector of a translation

9.7.10   We can, therefore, identify the transform 'A' to A' as:

$$\begin{pmatrix} 6 \\ 0 \end{pmatrix} + \begin{bmatrix} 0 & 3 & 5 & 4 & 1 \\ 0 & 2 & 2 & 4 & 3 \end{bmatrix}$$

in which I have used the curved brackets to indicate the COLUMN VECTOR which is to be added to each COLUMN of the matrix in turn to effect the transform.

9.7.11 We can now look at various other translations in Figure 9.10 and consider what COLUMN VECTOR ADDITION generates the translations A→A', A→B, A→C and A→D, respectively.

In each case you should be able to identify a single vector.

Can you see how the transform A→C is the same as (A→A' and then (+) A→B)? How does this compare with our earlier combining of the 'journey' vectors when moving round the outline of a shape?

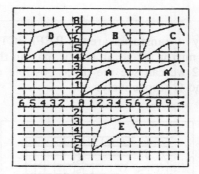

Figure 9.10 A variety of translations

9.7.12 It is also possible to combine *transforms* to produce particular effects. In the case of CAD software, for example, we often find the functions which allow us to present our basic designs *in perspective*. This involves both a translation and an enlargement or diminution of the original shape, a very simple example is shown here:

The original square has been reduced in size *and* translated '*left and up*' in order to help generate the simple perspective representation which we see in Figure 9.11.

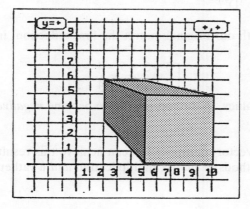

Figure 9.11 A perspective view of a square prism

9.7.13   What sequence of transforms would you use in order to achieve this? Is the sequence of operations important?

Try a number of such variations for yourself; sketching them in pencil on graph paper is the easiest way but *do* remember to set them out carefully and *do* set up the necessary matrix and the transform scalars and vectors to complete the operations on the *numeric* models.

There are many such combinations which we use in CAD and similar graphics software and we shall meet others.

9.8   Reflections

9.8.1   Yet another family of transforms are the REFLECTIONS, in which the shape is converted into its own mirror image, in the '*y*' axis, in the '*x*' axis, or *both*. In the illustration at Figure 9.12 we can see the three possible reflections of our original shape, 'A', into A', B and C respectively.

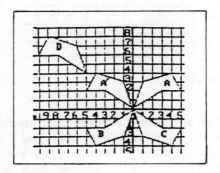

Figure 9.12   The possible reflections

9.8.2   What is shape 'D'?

9.8.3   Compared with our original matrix the first reflection, into A', is the matrix

$$\begin{bmatrix} 0 & -3 & -5 & -4 & -1 \\ 0 & 2 & 2 & 4 & 3 \end{bmatrix}$$

in which all the '*x*' values, in the top row, have become *negative*, (-0 does not exist, there is only 0).

9.8.4   No scalar or vector multiplication can achieve this result and it is time for us to meet both the methods of matrix multiplication *and* a rather unusual matrix.

9.9 Matrix Multiplication

9.9.1 In matrix multiplication there are a number of special rules and we need a little extra practice to become familiar with them.

The first rule is that the number of *columns* in the multiplier *must* match the number of *rows* in the multiplicand.

The second is rather more complicated and needs to be taken in stages. Taking the top row of the *multiplier* first, each of its elements, from left to right in turn, must multiply (in strict sequence) each element, from top to bottom, of the first column of the multiplicand matrix and the partial products must be totalled to give the *top left* cell of the product matrix.

The top row of the multiplier must then multiply the *second* column of the multiplicand, in the same fashion, to generate the *second column cell* in the *top row* of the product matrix.

This process continues until all the columns, from left to right, of the multiplicand, have been multiplied by the top row of the multiplier.

We then repeat the process with the *second* row of the multiplier, starting again with the leftmost column of the multiplicand, to furnish the *second row of the product matrix*.

This process continues until all row elements of the multiplier have multiplied all column elements of the multiplicand and we have a complete product matrix.

This is a rather involved process to explain and can be quite a tedious process to perform on a large matrix, manually, but a program segment to perform it is surprisingly brief. Let us explore the stages by using two matrices.

9.9.2 Under the first rule $\begin{bmatrix} 1 & 2 & 3 \\ 0 & 2 & 1 \end{bmatrix} * \begin{bmatrix} 2 & 1 & 0 \\ 1 & 4 & 3 \\ 2 & 1 & 5 \end{bmatrix}$ is legitimate but $\begin{bmatrix} 2 & 1 & 0 \\ 1 & 4 & 3 \\ 2 & 1 & 5 \end{bmatrix} * \begin{bmatrix} 1 & 2 & 3 \\ 0 & 2 & 1 \end{bmatrix}$ *is*

not since the columns in the multiplier outnumber the rows of the multiplicand.

9.9.3 If the multiplier is $\begin{bmatrix} 1 & 2 & 3 \\ 0 & 2 & 1 \end{bmatrix}$, designated as 'A', the multiplicand $\begin{bmatrix} 2 & 1 & 0 \\ 1 & 4 & 3 \\ 2 & 1 & 5 \end{bmatrix}$,

labelled 'B' and the product matrix 'P' let us 'step through' A*B = P, by the rules of matrix multiplication.

Step 1 is:

$\begin{bmatrix} 1 & 2 & 3 \\ . & . & . \end{bmatrix} * \begin{bmatrix} 2 & . & . \\ 1 & . & . \\ 2 & . & . \end{bmatrix}$ $= |1*2+2*1+3*2 = 10|$ to form the first cell in the product

matrix: $\begin{bmatrix} 10 & . & . \\ . & . & . \end{bmatrix}$

The next step is to multiply *column two by row one*

$$\begin{bmatrix} 1 & 2 & 3 \\ . & . & . \end{bmatrix} * \begin{bmatrix} . & 1 & . \\ . & 4 & . \\ . & 1 & . \end{bmatrix} = |1*1 + 2*4 + 3*1 = 12|$$ to form cell $p_{1,2}$ in our product

matrix, 'P':. $\begin{bmatrix} 10 & 12 & . \\ . & . & . \end{bmatrix}$

Next, row three performs: $\begin{bmatrix} 1 & 2 & 3 \\ . & . & . \end{bmatrix} * \begin{bmatrix} . & . & 0 \\ . & . & 3 \\ . & . & 5 \end{bmatrix} = |1*0 + 2*3 + 3*5 = 21|$ to form

cell $p_{1,3}$ in our product matrix: $\begin{bmatrix} 10 & 12 & 21 \\ . & . & . \end{bmatrix}$. We now move to the second row of our

multiplier and start with $\begin{bmatrix} . & . & . \\ 0 & 2 & 1 \end{bmatrix} * \begin{bmatrix} 2 & . & . \\ 1 & . & . \\ 2 & . & . \end{bmatrix} = |0*2 + 2*1 + 1*2 = 4|$ to form the

first cell in the second row of the product matrix:

$$\begin{bmatrix} 10 & 12 & 21 \\ 4 & . & . \end{bmatrix}$$ and continue the process until we have $P = \begin{bmatrix} 10 & 12 & 21 \\ 4 & 9 & 11 \end{bmatrix}$

9.9.4 As a pencil and paper process this is tedious for all but quite small matrices, but a program segment is relatively short. Here is one possible version in which merely changing the maxima for i, j and k will adapt the routine for any sizes of matrix, subject to the first rule, above. For brevity, I have dealt only with the 'core' activity and not declared the arrays, merely labelled them as in our worked example, above.

```
i=1
        DOWHILE i<3
        j=1
                DOWHILE j<4
                k=1
                        DOWHILE k<4
                        p_i,j ← p_i,j + (a_i,k * b_k,j)
                        k ← k+1
                        ENDDO
                j ← j+1
                ENDDO
        i ← 1+1
        ENDDO
END
```

9.9.5 BEWARE! Matrix multiplication is NOT COMMUTATIVE!

Try $\begin{bmatrix} 1 & 2 & 3 \\ 0 & 2 & 1 \end{bmatrix} * \begin{bmatrix} 1 & 0 \\ 4 & 3 \\ 1 & 5 \end{bmatrix}$ and then try $\begin{bmatrix} 1 & 0 \\ 4 & 3 \\ 1 & 5 \end{bmatrix} * \begin{bmatrix} 1 & 2 & 3 \\ 0 & 2 & 1 \end{bmatrix}$

Are the product matrices identical?

9.9.6 Let us now look at the special matrix which generates reflections. Because of the pattern of matrix multiplication we use many variations of the rather odd-looking

matrix $\begin{bmatrix} n & 0 \\ 0 & n \end{bmatrix}$ as a multiplier which *will multiply selectively*.

9.9.7 The rules are that this matrix must be a *square* with the same number of *columns* as there are *rows* in the matrix which is to be multiplied and that all cells not on the LEADING DIAGONAL shall be zero.

The leading diagonal is that from top left to bottom right and we replace the 'n' of the model, above, with whatever we need in order to achieve the required result.

9.9.8 Most importantly the IDENTITY MATRIX, with 1 in each cell of the leading diagonal and 0 in all others, will leave the *values unchanged* in any matrix which it multiplies.

In our present case, for example, $\begin{bmatrix} 1 & 0 \\ 0 & 1 \end{bmatrix} * \begin{bmatrix} 0 & 3 & 5 & 4 & 1 \\ 0 & 2 & 2 & 4 & 3 \end{bmatrix} = \begin{bmatrix} 0 & 3 & 5 & 4 & 1 \\ 0 & 2 & 2 & 4 & 3 \end{bmatrix}$

What happens if we make:

    a) the top left cell −1?
    b) the bottom right cell −1?
    c) *BOTH* cells on the leading diagonal −1?

What combination caused A→D in Figure 9.12?

9.9.9 This surprisingly simple device allows us to do some very sophisticated transforms and we will see other variations a little later.

9.10 Rotations

9.10.1 For the moment let us look at another major set of transforms, the ROTATIONS.

In Figure 9.13 we see our shape rotated to various extents until, finally, it can turn a little further from 'C' and become 'itself' again.

In other words a 360° rotation is achieved by the IDENTITY matrix once again.

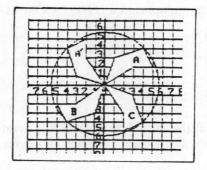

Figure 9.13 Just four of an infinite number of rotations

9.10.2 This time, however, we are concerned with the *amount of turn*, or the angle through which we turn our shape, as in the ROTATE function of many graphics programs.

9.10.3 The matrix which achieves the rotational transforms is $\begin{bmatrix} \cos\alpha & -\sin\alpha \\ \sin\alpha & \cos\alpha \end{bmatrix}$ where

α denotes the size of angle of *anti-clockwise* rotation which we seek.

9.10.4 It is perhaps not surprising that the transform matrix, for this purpose, contains FUNCTIONS OF THE ANGLE of turn. The operative matrix is one whose derivation need not concern us and the angular functions are normally available in most interpreters and compilers.

The only complication is that few such translators deal readily with the 60-based fraction system of degrees, minutes and seconds with which we customarily measure angles but use, instead, the decimal-based RADIAN MEASURE.

Perhaps we should pause for a moment and clarify what this is. Figure 9.14 may help.

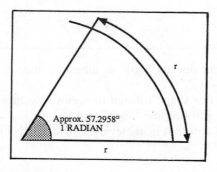

Approx. 57.2958°
1 RADIAN

Figure 9.14 One radian = 360°/2π

9.10.5 A distance, *equal to the RADIUS, along the circumference of the circle*, marks off an ARC which defines an angle at the centre of the circle of 1 RADIAN.

Since the circumference is $2\pi r$, or $(2 * \pi *$ the length of the radius) one *rad* $= 360°/2\pi$ which is approximately 57.2958°. This is a CONSTANT and to convert angles expressed in degrees to radian measure we divide by this constant.

Be sure to check carefully, in all computing work, whether *your* translator accepts angles in degrees or radians; you may need to write a conversion program module.

9.11   Glide and Stretch

9.11.1 When we perform both a rotation and a translation we describe it as a GLIDE.

There is, as with all our transforms, no real way of learning about such behaviour except by having plenty of practice and you should try as many examples as possible with different shapes, different amounts of turn and different combinations of transform.

9.11.2 So far all our transforms have left the original shape *unchanged* but let us look at other variants of the identity matrix and see what happens if we use values other than 1 in the leading diagonal.

$$\begin{bmatrix} 2 & 0 \\ 0 & 1 \end{bmatrix} * \begin{bmatrix} 0 & 3 & 5 & 4 & 1 \\ 0 & 2 & 2 & 4 & 3 \end{bmatrix} = \begin{bmatrix} 0 & 6 & 10 & 8 & 2 \\ 0 & 2 & 2 & 4 & 3 \end{bmatrix}$$

Plot this on your graph paper; how would you describe what has happened to our original shape?

9.11.3 This is a STRETCH. In this case we have stretched our shape along the '*x*' axis but left the '*y*' values unchanged. What happens if we make both cells on the leading diagonal 2; or top left 2 and bottom right $1\frac{1}{2}$? Clearly this is another transform which we need to investigate by experiment to be sure of what effect we shall produce with different values in the leading diagonal.

9.12   Shear

9.12.1 The final shape-changing set of transforms, for our present purpose, is achieved by yet another modification of the identity matrix. These perform an operation known as SHEAR of which Figure 9.15 gives us just one example.

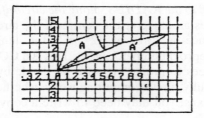

Figure 9.15 The shape distorted – but still 'true'

9.12.2 By using the transform matrix: $\begin{bmatrix} 1 & 2 \\ 0 & 1 \end{bmatrix}$ on our shape, 'A', we arrive at the product matrix, A', $\begin{bmatrix} 0 & 7 & 9 & 12 & 7 \\ 0 & 2 & 2 & 4 & 3 \end{bmatrix}$ and when this is plotted on our graph field we see the rather surprising result in Figure 9.15. It is as though a force pushed the shape, at the top-left corner, so that every part of the shape 'slides' sideways by a different amount.

These comparatively simple models, and the operations upon them, are equally well performed in three dimensions, by using ordered triples and three-row matrices. They *are* a little beyond the scope of this book but the principles are precisely the same and many of the most exciting developments in computer applications depend upon just such manipulations.

9.13 Conclusion

The graphic capabilities of the computer, at least in *vector* graphics, are really *numeric* capabilities relying on our understanding of the underlying mathematics in order to switch our thinking from graphic to numeric mode at will. The screen and the plotter are, ultimately, cartesian number spaces in common with our everyday graphs.

Exercises

Use graph paper to represent a VDU screen wherever necessary.

1) An aircraft is flying due South at 400 knots and there is a 50 knot wind blowing from the East. Plot the vectors and show the plane's actual track.

2) A boat sails 5km due South, then 7km North-East followed by 3km due North.

   a) Using the 'y' axis as North/South and a scale of 1cm : 1km plot its track and measure the distance and bearing it needs to sail in order to set back to its starting point.
   b) State this vector as an ordered pair.

3) A shape is identified by the matrix: $S = \begin{bmatrix} 1 & 5 & 12 & 0 \\ -2 & 1 & 1 & 5 & 2 \end{bmatrix}$

   Plot the shape on a graph

4) In each of the following operate on 'S', as in Question 3), and plot the change on graph paper; (OR, if you have a suitable computer system, write the appropriate short program and test it).

   a) What is the result of $\begin{pmatrix} 2 \\ 3 \end{pmatrix} + S$?

   b) Calculate $\begin{bmatrix} -1 & 0 \\ 0 & 1 \end{bmatrix} * S$

      i) Plot the resultant shape.
      ii) What transform has been effected?

   c) What is the effect on the original shape of 1.5 * S?

   d) $\begin{bmatrix} 0 & -1 \\ 1 & 0 \end{bmatrix} * S$

   e) $\begin{bmatrix} -0 & 1 \\ 1 & -0 \end{bmatrix} * \left(\begin{pmatrix} 1 \\ 2 \end{pmatrix} + S \right)$

   f) $\begin{bmatrix} 1 & 1.5 \\ 0 & 1 \end{bmatrix} * \left(\begin{bmatrix} -1 & 0 \\ 0 & 1 \end{bmatrix} * S \right)$

5) Write and test a pseudo-code segment to perform each of the tasks in Question 4.

169

10
Data Structures

Objectives

After working through this chapter you should be able to:
- understand the relevance and operation of arrays, lists, stacks, queues, trees and similar elementary data structures;
- program their loading, searching and ordering;
- describe how networks are stored and manipulated as arrays;
- understand and use tables with pointers and back-pointers;
- select the structure which is appropriate to the task.

10.1 Organising Data

10.1.1 Whilst the computer is capable of very fast processing, it wholly lacks imagination or any ability to 'think' for itself. One consequence of this is that it is very difficult to write programs to make much sense of data which are presented or stored haphazardly. It is also true that the most *efficient* processing can be attained when data is organised in ways which suit the machine particularly well. Once again, mathematics offers ready-made solutions to such problems.

10.1.2 The first thing which we demand of good storage and processing is that the data be organised according to an efficient *pattern* so that we do not need a separate memory address for each data item. Ideally, the nature of the pattern should be defined *simply*, either within the program or by means of a *header* within the data set itself, perhaps as the first record, or as a 'flag' to that record.

10.1.3 The second requirement is that we should be able to find any *individual* piece of data quickly. Some arrangements lend themselves to faster searches than others.

10.1.4 Thirdly, we need to be able to insert, exchange or delete individual data items readily.

10.1.5 Acceptable *practical* DATA STRUCTURES must satisfy these criteria and it is essentially such mechanisms which we shall begin to study in this chapter. We shall look at some of the fundamental data structures and examine how they are used and what elementary manipulations may be performed on them.

10.2 Matrices Again

10.2.1   Since so much of the data we are handling is *numeric*, it is not surprising that the MATRIX or ARRAY is one of the commonest data structures. We have already looked at the basic principles of matrices and seen some applications. For most purposes we commonly process arrays by 'nested loops' in our programming.

10.2.2   To load the data into an array, for example, we might first *declare* the array by a) defining the data type and b) *dimensioning the array*. We may for example, make two program statements such as 'DEFINT A', to declare A as an integer, and 'DIM A(2,5)', to dimension the array A as a 2-row, 5-column matrix.

10.2.3   We are not tied to *two* dimensions but may create one-dimensional arrays (vectors) or define several dimensions; the computer is not limited in this respect.

10.2.4   To input a data set into the matrix, A, we might then use a segment of 'nested loop' code rather like this:

```
i=1, j=1;
DOWHILE i<3
        DOWHILE j<6
            READ A_{i,j}
            j ← j+1
        ENDDO
        j ← 1, i ← i+1
ENDDO
END
```

which will read successive inputs into $a_{1,1}$ $a_{1,2}$, ... $a_{1,5}$ in the first 'pass' of the 'i' loop, involving *five* passes of the 'j' loop. On the second pass of the 'i' loop cells $a_{2,1}$, $a_{2,2}$, ... $a_{2,5}$ will be 'loaded'.

10.2.5   The same program segment can be adapted to perform any operation normally performed on an array. Given the speed of processors, this is a perfectly adequate mechanism for the vast majority of array processing but, for some highly specialised purposes, ARRAY PROCESSORS may be used. In machines which are so configured each cell, of perhaps an 8×8 array, is allocated its own dedicated processor so that all cells may be processed simultaneously (rather than serially, as in our nested loop structure).

Data sets which are larger than 8×8 can then be sub-divided into a sequence of arrays of that size for processing.

10.2.6   It is common for spreadsheet systems to permit any two or more sub-sections of the sheet to be defined as arrays and the normal operations of add, subtract, multiply and divide can be called by fixed commands in the spreadsheet meta-language or macros. Since the commutative and inverse laws do not operate in the same way for matrices *we* must specify not only the arrays to be used and an output array but, most importantly, the *sequence* of arrays in the calculation.

10.2.7   The common operations on a matrix, some of which we have already met in other contexts, are:

1) It may have a constant quantity *or* a constant vector, added to it or subtracted from it.
2) It may be multiplied, or divided, by a (constant) *scalar*.
3) It may have another matrix, *of the same dimensions*, added to it or subtracted from it by *cell-by-cell mapping*.
4) It may, if we specify the procedure clearly, be subjected to a *cell-by-cell* multiplication by another matrix of *identical dimensions*. This is *not true* matrix multiplication and cannot usually be achieved by calling such a function, even within a spreadsheet. (In the latter case performing a formula multiplication of cells (1,1) and then using the COPY function is often the answer but within a specific program the nested loop construct will serve.)
5) It may be multiplied by a *row vector* which has the same number of elements as the original matrix has rows or it may multiply a *column vector* which has the same number of rows as the original matrix has columns.
6) It may be multiplied, selectively, by a *square matrix* of which the number of columns matches the number of rows in the multiplicand.
7) It may be multiplied (in *true* matrix multiplication, sometimes described as 'row-by-column'), by a matrix, *the number of whose columns match the number of rows of the multiplicand matrix*. For example a 4×6 matrix, in 'full' matrix multiplication, may be multiplied by any 'n'×4 matrix where $1 \leqslant n \leqslant 6$.

10.3   An Array Example

10.3.1   Let us take a practical example and consider in turn, each of these processes which we have not previously met. We first need the 'narrative' of the problem which we must model:

"A manufacturing company, in factories at Andover and Bristol respectively, employs the following numbers of people in 4 grades for each of 6 job-categories and at different monthly pay scales. (For simplicity the data are given in the tables.) There are a number of questions to which management need answers and other such questions will arise in the future."

10.3.2   Table 10.1 shows the number of staff in each of the two factories set out in rows and columns and this already constitutes a matrix of a sort, but it is not yet very suitable for manipulations. For that we must simplify yet further.

Table 10.1

GRADE	ANDOVER						BRISTOL					
	CATEGORY						CATEGORY					
	A	B	C	D	E	F	A	B	C	D	E	F
1	0	2	5	16	0	0	0	3	18	0	0	1
2	2	12	12	27	5	1	1	4	12	18	2	0
3	0	0	24	15	3	2	0	0	16	14	5	1
4	0	0	10	9	2	1	0	0	8	6	1	0

10.3.3 In the Table 10.2, we have the monthly pay *rates* set out and so far as the wages element of costs is concerned these three sets of data, when organised into the appropriate matrices, will allow us to derive a good deal of further information by using, at different stages, some of the operations listed as 1) to 7), in Section 10.2.7.

10.3.4 The first step in modelling the situation is to create true matrices from the known data with all rows and columns labelled appropriately.

It does not matter particularly which items we put into the rows and which into the columns so long as, thereafter, we are wholly consistent.

Table 10.2

SCALE	WAGE RATES $00					
	CATEGORY					
	A	B	C	D	E	F
1	1	1	1	1	2	2
2	2	2	3	3	4	5
3	3	3	4	4	6	8
4	8	8	9	9	10	12

If A=Andover, B=Bristol and W=Wage rates then:

$$A=\begin{bmatrix} 0 & 2 & 5 & 16 & 0 & 0 \\ 2 & 12 & 15 & 27 & 5 & 1 \\ 0 & 0 & 24 & 15 & 3 & 2 \\ 0 & 0 & 10 & 9 & 2 & 1 \end{bmatrix} \quad B=\begin{bmatrix} 0 & 3 & 18 & 0 & 0 & 1 \\ 1 & 4 & 12 & 18 & 2 & 0 \\ 0 & 0 & 16 & 14 & 5 & 1 \\ 0 & 0 & 8 & 6 & 1 & 0 \end{bmatrix} \quad W=\begin{bmatrix} 1 & 1 & 1 & 1 & 2 & 2 \\ 2 & 2 & 3 & 3 & 4 & 5 \\ 3 & 3 & 4 & 4 & 6 & 8 \\ 8 & 8 & 9 & 9 & 10 & 12 \end{bmatrix}$$

contain all the known data and may now be manipulated in order to extract information.

10.3.5 QUESTION: How many people *in all* are employed in *each* of the 24 sub-groups?

10.3.6 METHOD: Add the two 'numbers' matrices, A and B, cell-by-cell, to give 'T', the totals by category/grade.

$$\begin{bmatrix} 0 & 2 & 5 & 16 & 0 & 0 \\ 2 & 12 & 15 & 27 & 5 & 1 \\ 0 & 0 & 24 & 15 & 3 & 2 \\ 0 & 0 & 10 & 9 & 2 & 1 \end{bmatrix} + \begin{bmatrix} 0 & 3 & 18 & 0 & 0 & 1 \\ 1 & 4 & 12 & 18 & 2 & 0 \\ 0 & 0 & 16 & 14 & 5 & 1 \\ 0 & 0 & 8 & 6 & 1 & 0 \end{bmatrix} = \begin{bmatrix} 0 & 5 & 23 & 16 & 0 & 1 \\ 3 & 16 & 27 & 45 & 7 & 1 \\ 0 & 0 & 40 & 29 & 8 & 3 \\ 0 & 0 & 18 & 15 & 3 & 1 \end{bmatrix}$$

is what A+B=T involves and is arrived at by the

$$a_{1,1} + b_{1,1} = t_{1,1} \text{ until } a_{4,6} + b_{4,6} = t_{4,6}$$

sequence in a 'DOWHILE i<5; DOWHILE j<7' nested loop. Try to write the pseudo-code routine for this, remembering to declare and initialise all necessary matrices and variables.

10.3.7 QUESTION: How many people are employed, *in total*, at 1) Andover; 2) Bristol?

10.3.8 METHOD: *Step 1* Multiply T by the appropriate UNIT ROW VECTOR. (A UNIT ROW VECTOR is a single row matrix which has a 1 to match each ROW of th multiplicand matrix. The subsequent 'multiply by 1' process is using the law of identity to select and total cells without changing the individual values.):

$$(1\ 1\ 1\ 1) * \begin{bmatrix} 0 & 2 & 5 & 16 & 0 & 0 \\ 2 & 12 & 15 & 27 & 5 & 1 \\ 0 & 0 & 24 & 15 & 3 & 2 \\ 0 & 0 & 10 & 9 & 2 & 1 \end{bmatrix} = (2\ 14\ 54\ 67\ 10\ 4)$$

Step 2 Multiply the UNIT COLUMN VECTOR (which has a 1 to match each column of the multiplier ROW VECTOR) by the ROW VECTOR generated in Step 1.

$$(2\ 14\ 54\ 67\ 10\ 4) * \begin{pmatrix} 1 \\ 1 \\ 1 \\ 1 \\ 1 \\ 1 \end{pmatrix} = 151$$

Can you a) perform the same calculation for Bristol, b) for both factories *in total*, c) write the pseudo-code segment which will perform this task?

10.3.9   QUESTION: What is the total wages bill?

10.3.10   METHOD: *Step 1* Perform the cell-by-cell addition, A+B to find the 'total employees' matrix, 'T'.

Step 2 Next, do the *cell-by-cell* multiplication of W*T to generate the matrix which will contain (by category and grade) all wages 'C'osts.

$$C = \begin{bmatrix} 0 & 5 & 23 & 16 & 0 & 2 \\ 6 & 32 & 81 & 135 & 28 & 5 \\ 0 & 0 & 160 & 116 & 48 & 24 \\ 0 & 32 & 162 & 135 & 30 & 12 \end{bmatrix}$$

Step 3 Multiply C by the appropriate unit row matrix to total the columns. (This technique may be useful for answering additional questions. Can you suggest such a question?)

Step 4 Multiply the appropriate unit column vector by the product of step 3 to generate the overall total. If you have followed steps 3 and 4 correctly your answer should be $102\ 000.

10.3.11   QUESTION: What is the total wage bill for employees in categories C to E, inclusive, at grades 2 and 3, of matrix C?

$$\textbf{10.3.12}\quad \text{METHOD } (0\ 1\ 1\ 0) * \begin{bmatrix} 0 & 5 & 23 & 16 & 0 & 2 \\ 6 & 32 & 81 & 135 & 28 & 5 \\ 0 & 0 & 160 & 116 & 48 & 24 \\ 0 & 0 & 162 & 135 & 30 & 12 \end{bmatrix} * \begin{pmatrix} 0 \\ 1 \\ 1 \\ 1 \\ 1 \\ 0 \end{pmatrix}$$

It is important that you should work this example and others like it so that you can see clearly the *selective* operation of the unit vector, row *or* column, in which some of the 1s have been replaced by 0s, putting into effect the law of multiplication by zero to eliminate unwanted partial products in the matrix multiplication.

10.3.13   QUESTION: What will be the effect on the total wage bill if all grade 1 staff are given wage increases of 5%, grade 2, 6%; grade 3, 4% and grade 4, 8%?

10.3.14   METHOD: *Step 1* Since multiplying by 1.05, for example, will increase an amount by 5% we use a *modified* identity matrix (as we saw in the previous chapter for selective enlargement of shapes) which will first multiply each *row* by its appropriate scalar:

$$\begin{bmatrix} 1.05 & 0 & 0 & 0 \\ 0 & 1.06 & 0 & 0 \\ 0 & 0 & 1.04 & 0 \\ 0 & 0 & 0 & 1.08 \end{bmatrix} * C = \begin{bmatrix} 0 & 5 & 23 & 16 & 0 & 2 \\ 6 & 32 & 81 & 135 & 28 & 5 \\ 0 & 0 & 160 & 116 & 48 & 24 \\ 0 & 0 & 162 & 135 & 30 & 12 \end{bmatrix}$$

and find the *new* total wages matrix C'. Work this yourself before moving to Step 2 and again, note the effect of the multiplications by zero.

Step 2 Find the column totals by multiplying the new matrix by the appropriate unit row vector.

Step 3 Find the overall total by multiplying the appropriate unit column vector by the product of Step 2.

How would you evaluate the total increase? If you answer is 5.94% to 2D, your intermediate steps were probably sound.

10.4   Arrays and Spreadsheets

10.4.1   It is especially easy, with computerised manipulation of arrays, to perform 'WHAT IF?' modelling of real situations. In our previous example, 'What if the grade 3 people had a 6% increase?' means only changing that one cell in the transform matrix and re-running the short program. In such ways, using an algorithmic structure-handling process or processes we can nevertheless develop and use genuinely 'try it and see', heuristic, models.

10.4.2   All such capabilities are built into any reasonable spreadsheet system where we can identify any sub-section of the spreadsheet (itself an array) as a separate matrix, by identifying the RANGE of cells occupied by the selected matrix, on which we may operate by any of the techniques we have discussed here. To avoid a lot of pencil-and-paper working, with the risk of making mistakes, a spreadsheet package was used to perform all the calculations needed in exploring our examples.

10.4.3 Try out on a spreadsheet, for yourself, the various operations we have performed, above. You will probably be asked to identify, by *range*, matrix 1, the multiplier, then matrix 2, the multiplicand and finally the *destination* of the product matrix, as its top left-hand cell. You should try full multiplication, as well as multiplication:

  – by a row vector; by a unit row vector; by a selective row vector;

  – of a column vector; of a unit column vector; of a selective column vector.

10.5   Multi-dimensional Arrays

10.5.1 For practical purposes pencil-and-paper modelling is limited to fairly small two-dimensional arrays but spreadsheets offer us much more power than this.

10.5.2 However, it is undoubtedly within a program that we can develop the greatest power in manipulating arrays. For instance, we may find it difficult even to envisage three-dimensional data structures but the computer has no experience of the real world to inhibit it and can deal as readily with *six*-dimensional arrays if they suit our purpose. This may sound a little strange at first but if we are dealing with three-dimensional movement, of an aircraft let us say, we are also dealing with both speed and time and a five-dimensional data array may be extremely useful.
Both the structure and the methods of manipulating data within the structure have many applications and are of very great importance.

10.6   String Arrays

10.6.1 The computer gives us yet more power because, although we cannot perform arithmetic operations on them, we *can* store strings in arrays, as well as numeric data, and may well be able at least to 'concatenate' the elements of one array with the elements of another. The functions available will depend upon the programming language and the individual interpreter software.

10.6.2 Unless our strings are of very varied lengths the array is a very useful structure for their storage and handling, especially in main memory because of its economical use of space.

10.7 Array Addresses

10.7.1 Within the computer's memory, when we define an array, it is given a start address and a total length, usually in bytes. The address label is transparent to the user but many translators allow us to determine it if we wish.

10.7.2 If an array is defined as 4×7 integer, and the computer stores integers in 16 bit, (2 byte), form, we can readily visualise the array. From the starting address the control unit will allocate a nominal 4*7*2 = 56 bytes of memory. Of this, since it is identified as 4×7, the first quarter of the space, 14 bytes, will contain the cells, in order, of the top row of the matrix. The next quarter will contain the second row, and so on.

10.7.3 We need know only the *starting address* where the array is stored; all else is relative to this and we speak of 'relative addressing'. We would load our data into each of their appropriate cells by some variation of a nested loop 'read' program segment.

10.8 Lists

10.8.1 When we look at a *real* data item, say one record in an invoices file, we often find that a single matrix cannot hold all the elements because they are of different data types. Such a record will need at least four fields, INVOICE NUMBER: CUSTOMER NUMBER: DATE: AMOUNT.

The first item might be alphanumeric, the second an integer, the third a date string of the *fixed* dd/mm/yy type whilst the fourth may be a real number.

10.8.2 We could not define a four row matrix to hold four different variable types such as these and would need to declare *a different one-dimensional array* for each different data type.

10.8.3 Any such array or vector may also be thought of as a LIST and lists may be INDEPENDENT or LINKED. In the example shown in Figure 10.1, all four elements of the data record must stay together. This is not always easy since we may well, for different reports, need to *order or 'sort'* each list individually.

10.8.4 In this case, as in Figure 10.2, the individual elements will no longer necessarily be anywhere near their original positions and our four lists would need to be linked, laterally, so that any individual invoice number 'points' to the position in the customer number list where the second element of the whole record is to be found.

Similarly the customer number should point to the date and the date to the amount so that, under all circumstances, entering the invoice number will automatically allow us to retrieve all the other elements of the corresponding record.

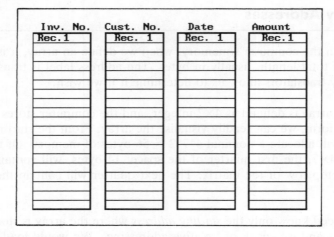

Figure 10.1   Four lists which must stay 'linked'

10.8.5   In practice we might well want all pointers to operate in both directions so that, for instance, we could retrieve all invoices billed on a given date, or for a particular customer number, or for a given range of values.

10.8.6   The particular problem is that the *unique identifier* for each record is the invoice number, so that it is only this which establishes the combination of lists as *a set*.

10.8.7   At all costs we must preserve the mapping of the invoice number list onto each of the other three or we shall finish up with one set and three jumbled 'heaps'.

10.8.8   The easiest way, in our matrix techniques, of preserving the linking identity of each element of a number of lists simultaneously is to 'flag' each item by its *subscript* at the time of original entry, and to append this number to the data element as it is stored.

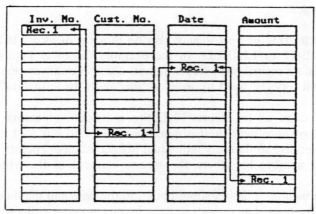

Figure 10.2   The lists after ordering

10.8.9 Such a device must be planned and provided for from the very beginning of program design and there is a strong case for including all such pointers in the data dictionary and all other documentation. The pointers may indicate one direction only, and a one-to-one mapping, or may be required to indicate a one-to-many mapping.

10.9 System Lists

10.9.1 Program location in main memory and its control during execution is often by means of linked lists and here is a sketch of how this might be achieved. Any single process or command will 'call' a particular segment of program code. Within this, in turn, may well be one or more calls to sub-routines.

10.9.2 If each of the sub-structures of a single program, or the operating system, is brought into memory as a list in a preset position we can, by setting suitable pointers, both economise on storage and speed up the running of the program.

10.9.3 We often achieve this by recording or 'poking' the relevant memory addresses into another list dedicated to the pointers themselves (see Figure 10.3).

10.9.4 Yet another use of lists is in holding in memory the various interrupt vectors which control the whole computer configuration while it is running.

10.9.5 Some understanding of the use of lists, their setting-up and operation does help us design better programs and understand what is going on when we read the programs of others.

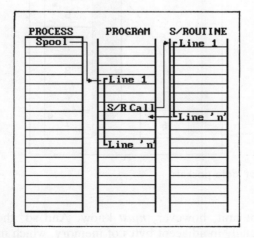

Figure 10.3 Program lists with pointers

10.10   Stacks

10.10.1   There are different ways of establishing and handling lists. If we venture to read a manual on low-level programming, in assembler for example, we shall quickly meet the word 'stack' just as, in a job control language, or when managing a print spooler, we meet the word 'queue'. This applies to all levels of computing practice, not merely to main-frame installations.

There is nothing to fear in these words since they, together with several others, are only labels for variants of one-dimensional arrays. What is different is the ways in which data are manipulated into and out of the arrays.

10.10.2   Any single item of data is, in strict mathematical language, a 'datum' and we shall start to use this word rather than repeating phrases like 'data item'. The other jargon words are 'push', for entering a datum into an array, and 'pop' for retrieving a datum.

10.10.3   If we pop and push data on a 'LAST IN FIRST OUT' system we refer to such an array as a *stack*.

10.10.4   Figure 10.4 shows a graphic model of such a stack and there are a number of things to note. If we have allocated a stack for 16 data and, as yet, we have included only 5 of those data, we shall neither know nor care exactly whereabouts in the stack they are.

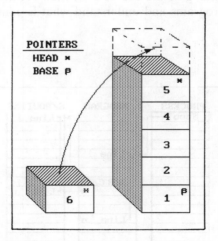

Figure 10.4   A stack of 'data packets'

10.10.5   The control unit, however, *must* know. And so, though invisible to us, 'pointers' are set, usually in adjacent bytes of memory, which indicate exactly where both the first *and* the last data to be pushed into the stack are to be found.

10.10.6   If, as suggested in Figure 10.4, another datum is then pushed into the stack it will be set 'on top of' datum 5 and the stack 'top' pointer will be reset to indicate datum number 6.

10.10.7   Equally if datum 6 is 'popped' *from* the stack, as the last in and therefore the first out, the 'top' pointer will be reset to indicate datum 5. This, like so many things in computing, is lengthy to explain but such data movements and the amendment of the corresponding pointers happen very frequently and at very high speed without the user, in general, ever being aware of it.

10.10.8   If we have sent a number of jobs to the print-spooler, we might wish them to be processed in the order in which they were prepared.

This, like a queue at a shop, involves creating a FIRST-IN-FIRST-OUT system, a very different operation of a stack.

When we represent this in similar fashion to our original stack, it is a little clumsy since, if we wish to push a datum into the stack we must first move the latter upwards in order to make space for the new datum to be added to the *bottom* of the stack. (The pointers in Figure 10.5 are shown solid, for the present condition of the stack, and in outline for the changes which will take place to both top and base pointers if we push a new datum in or pop the first out.)

Because of the awkwardness implied by this vertical sketch we often show a queue in a diagram as a *horizontal*, rather than a vertical, array but it is still commonly referred to as a PUSHUP STACK or LIST.

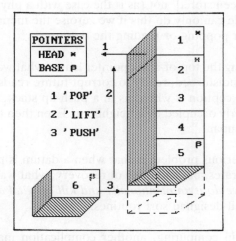

Figure 10.5   The clumsiness of a queue 'stack'

10.11   Deques and Wrap-arounds

10.11.1   For some purposes a queue may have data pushed and popped at both ends and you may find references to this as a DEQUE, or double-ended-queue.

10.11.2   Our final example of a stack is a WRAP-AROUND-STACK which may be seen as a kind of data merry-go-round in which a stack of fixed length is arranged so that the last datum cannot be pushed out of the end if the pushup or pushdown goes too far but will, instead, transfer to the *first* datum position and so not be lost.

10.11.3   All such data structures are vital to the most fundamental operations of the machine in its *internal* handling of data and, in particular, for the storage and line-by-line running of our programs.

Mostly these functions are transparent to us but job-scheduling, print spooling and many other operating system activities are intimately connected.

They are, however, also extremely useful in applications programming. We can and should, especially if we have ambitions to be systems programmers, simulate the operation of stacks and queues by short program segments in high-level languages so that we may become familiar with all the principles involved.

10.12   Corruption by the Machine

10.12.1   There are two very important little quirks of the machine which we must always remember. It does *not* 'clean up' after itself nor does it make sure that a location is empty before it pushes a data item into it!

In practice, when we pop or read a datum it is merely *copied* from the memory 'cells' in which it has been stored, not (as is the case with a physical *parcel* in a stack) physically removed. We can only do this if we zeroise the memory location, to 'clear out' the old item, after popping or reading the datum.

10.12.2   Even re-setting the top-of-stack pointer leaves what was stored in the former top-of-stack location undisturbed, ready to corrupt future reads of that location if we are not careful. The exception is when, as in a push-up stack, each datum is moved into the position formerly occupied by a neighbour. Even then the 'last' cell will often contain its original occupant.

10.12.3   An equally serious problem is that when a datum is pushed or written into a memory cell it obliterates, beyond hope of recovery, what was there before. This is not a problem *unless we do it unintentionally and kill off vital data*. This is especially likely to happen with ill-designed sort routines.

10.12.4   As so often in computing, another complication may arise when we are pushing or writing string variables. If we push a short string, say "Invoice", into a location which previously held a longer string, say "calendar", we *may* find, when we next retrieve the string, that it has become "Invoicer".

Can you see how? We need to check whether or not a string is automatically 'padded out' with spaces to overlay the whole of any previous string. Since it is usually *we* who define the individual string datum length to be allocated in any string array, we must also be careful that we don't finish up with "Invoi".

10.13 Ordering and Searching

10.13.1 The processes with which we are involved quite frequently are first the ORDERING of arrays, lists and stacks, and second, SEARCHING for individual data.

10.13.2 WARNING! One thing which we should never attempt to do, since each cell is directly related to all others in the same column and the same row, is to *order* a two- or more-dimensional matrix. Position, in a matrix, is *fixed* and is usually specified by the LABELS which we attach to rows and columns.

Only *lists* of the various kinds may safely be ordered and even then we must, before doing so, very carefully check any mappings which there may be on to other lists! If two lists *are mapped on to each other* or linked in any such indirect way, extra care is needed to ensure that all necessary pointers are generated or that all moves of a datum in the one list are matched by corresponding moves in the other(s).

10.13.3 Ordering data arrays is usually described in computer jargon as 'sorting' and since this word really refers rather to 'classifying' or putting into categories, it does cause some confusion. However, the jargon is well established and we must use it.

10.13.4 Essentially all SORTS are variations on putting the elements of an array into either *ascending* or *descending* order of either *magnitude* or *alphabetical order*. Since letters and all other symbols are held in the machine in numeric codes *all* sorts are, albeit transparent to the user, orderings by *numeric magnitude*. For *ascending* order sorts, they are all variations on:

If $a_i \leq a_{i+1}$ then LEAVE else SWAP THEIR POSITIONS.

10.13.5 For *descending* order sorts we replace the '$\leq$' sign by the '$\geq$'. We compare each pair of numbers in the array until the sort is accomplished. This is invariably done by finding DIFFERENCES. If $A - B < 0$ then $B > A$, and so on: all *ordering* relationships must be within the logical set $A>B$; $A<B$; $A=B$.

10.13.6 For strings we insert an additional nested loop to make additional 'compares', letter-by-letter, if the two first letters are 'equal'.

10.13.7 In a simple 'ripple' sort we compare the first data item with each of the others in turn until we have the largest in a_n. Next we compare a_2 with each of the remaining cells, and so on. For an array of 'n' items, with a total of n-1 'passes':

```
i=1;
DO UNTIL i = n-1
j = i+1
        DO UNTIL j = n
        If a_i > a_j DO SWAP
        j=j+1 ENDDO
i=i+1
ENDDO
END
```

10.13.8 The swap routine is a little complicated by the effect noted above, that pushing a datum into a location obliterates what was there before.

We see, in state i) of Figure 10.6 the situation where n_1 and n_2 must be swapped in an ascending order sort.

Moving n_1 into the cell occupied by n_2, as in frame ii) of the diagram obliterates the '5' and leaves a '3' in both cells.

Reversing the sequence, as in frame iii) simply loses the '3'.

10.13.9 What we must do is to declare a spare 'sort' cell, or cells if we are moving the contents of other lists at the same time, so that we can perform a 'rotary' swap, as in frame iv).

Here in move 1 the '5' moves to the 'sort' cell; in move 2 the '3' replaces the '5' as n_2 leaving the way clear for move 3, indicated by the dotted line, in which the '5' is moved from the sort cell into n_1 to complete the swap.

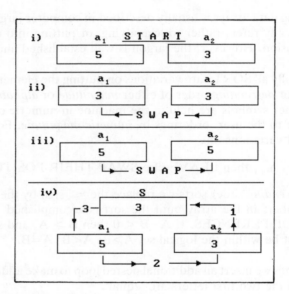

Figure 10.6 The place of the 'sort' cell

10.13.10 For short lists the ripple sort is very simple to program and relatively easy to understand. It does, however, need n-1 passes and, on average, ½n swaps and there are many other sorts which are faster and more powerful. All have a strict mathematical foundation, as do the SEARCHES, by which we find a specific datum in an array. In order to search we must first know what we seek and then compare that 'key' with each item in a list until we find a perfect match.

10.13.11 In an 'n'-dimensional matrix if 'n' > 1 we can commonly search by a nested loop routine which compares our key with each cell in turn, row-by-column. This is because the data are unsorted.

10.14 The Binary Chop

10.14.1 In an ordered list, or linked lists, the process is much faster by a simple BINARY CHOP search where, on each pass, half the list is eliminated. Suppose that we have a list of 600 items to search. We start by testing our key against item number 300 and, if this is larger than the key we know that the 'top' 300 items can be eliminated from our search.

On pass number two we compare with the 'middle' item, No.150. Let us say that we find this to be smaller than the key – the lower half of the remaining list may be ignored.

Test item number 225 comes next and then either 262 or 188, as the case may be.

10.14.2 Can you continue the sequence if the key number is 288, for example? Why is this called a 'binary' chop? Can you write, and prove by desk-top running, a segment of pseudo-code which will perform this search? What is the minimum number of passes needed to find a single item in a list of 1000? 10,000? 1 million?

Once again we have examined the simplest such routine. There are many others which mathematicians have devised in order to search larger and larger lists, whether fully ordered or not. This is still the subject of research and new routines *are* still being developed, but this happens comparatively rarely today since so many have already been discovered.

10.15 Trees

10.15.1 Quite often we have data in a list in a sequence which we need to preserve but which we need both to search quickly and to output in *ordered* form at will. A DATA TREE structure lends itself to such management of data, in particular the BINARY TREE.

We are all familiar with the growth pattern of a tree. From the root comes the trunk, and from the trunk come limbs, from which in turn come branches, which put out twigs from which grow twiglets and ultimately leaves or fruit.

10.15.2 In the much-simplified representation of a tree, in Figure 10.7, I have shown TWO limbs growing from the trunk, at level 2, TWO branches growing from each limb, level 3, and so on.

10.15.3 By restricting our model to two products of each 'split' we achieve two things. First, we can make each choice of path a yes/no decision, which fits with the binary logic of the computer.

Secondly, we can bring into usable form a pattern of growth which is, in nature, too complicated and unpredictable to lend itself to clear modelling.

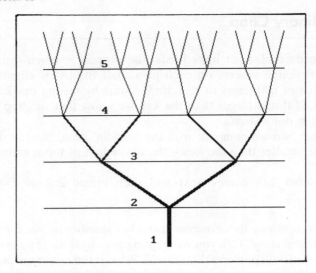

Figure 10.7   A simplified (binary) tree

10.15.4   As always when we meet a new modelling technique, there is some new language to master and here we find people in the industry speaking of a FOREST when our data structures include several trees, the ROOT, or basic element of the tree from which all others grow, right up to the LEAVES, which may be the last LAYER or GENERATION of the development of the tree.

10.15.5   The BRANCHES or PATHS are shown in Figure 10.8 along with the NODES. These are the key points in any such diagram and have been emphasised here.

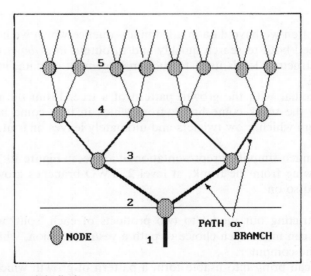

Figure 10.8   The tree labelled

10.15.6   At the *nodes* is where data are stored and where BRANCHING may take place in any of TWO directions.Only this last point determines a BINARY TREE since in any other model, depending on its complexity, there may be more than two paths joining or separating at any node.

10.15.7   We are also familiar with the idea of a tree as a data structure in connection with our own families, since a FAMILY TREE is one way of making clear all the relationships through several generations.

Such a tree, unlike what we have seen above, 'grows' downwards from the farthest traceable ancestor to the present youngest generation of children as shown in Figure 10.9.

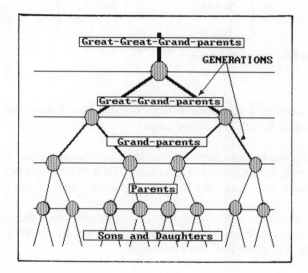

Figure 10.9   The family tree model

10.15.8   All members of the family are DESCENDANTS of the *root* from which they have sprung through several GENERATIONS. In the example here, I have gone to only five generations and kept to a binary model for the sake of simplicity.

This is quite unlike most *real* families where multiple branching would be much more likely and the tree which results may be extremely complicated.

10.15.9   Nevertheless *this* is the model which is most commonly used to illustrate our use of trees, growing 'down' rather than up, with successive layers referred to as *descendants*. Many writers do, quite haphazardly, mix the botanical language of trees and the 'descendants' language of *family* trees but you may find it simpler to stick to just the one pattern until the subject is completely familiar to you.

10.16   'Growing' a Data Tree

10.16.1   Let us take a list of data and set about developing a tree from it. The most important point about the technique is that we use this special diagram (shown in Figure 10.10) at a node to hold a single datum and to give us the essential information to describe the tree completely in due course.

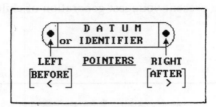

Figure 10.10   The fully annotated node

10.16.2   The symbol is a rectangle which contains the datum and on each end of which is a semi-circle which is initially empty.

10.16.3   If a subsequent data item is less than, or comes earlier in sequence than, the node concerned, we indicate the necessary BRANCHING (TO THE) LEFT by putting a small circle or a large dot into the left semi-circle of the node.

10.16.4   If a subsequent item comes *later*, or is larger, than the current datum we mark the right semi-circle in the same way. Since we are dealing exclusively with *binary branchings* this covers all the possibilities.

10.16.5   The circles or dots are like signposts pointing to left and/or right so we call them the RIGHT and LEFT POINTERS. If either or both possible branches remain unused we leave the appropriate semi-circle(s) empty. If both semi-circles are empty, then, we have reached a dead-end on that particular branching path.

10.16.6   We will follow the process with the data list:

Malaysia; Namibia; Mauritius; Bahrain; Hong Kong;
Pakistan; Sri Lanka; Botswana; Singapore and India.

10.16.7   When these are 'loaded' into our tree, in that order, the *root* will be Malaysia. You can see this in Figure 10.11.

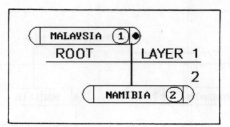

Figure 10.11 The first descendant

10.16.8 Item two, Namibia, comes *after* Malaysia in alphabetical order and will branch to the right, or, in the jargon, be a RIGHT DESCENDANT and we SET the right pointer. NOTE: it is not always done in practice but I have *numbered* the nodes, in small circles alongside the datum, according to their places in the original sequence. This is only to help us keep track of what we are doing and is a good practice to follow until you are completely confident in developing trees.

10.16.9 Continuing down our list, we come next to Mauritius, which is alphabetically before Namibia, but *after* Malaysia, hence it becomes a *left-descendant* of Namibia. The first three nodes, the root and its first two descendants, will look like Figure 10.12.

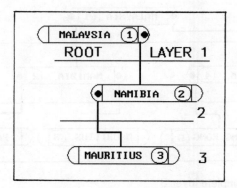

Figure 10.12 The first three nodes

10.16.10 Bahrain is *ahead* of Malaysia in alphabetical order and will become the first LEFT DESCENDANT of the former on layer 2. This is shown in Figure 10.13.

The diagram is beginning to get a little crowded and you will notice that the branch from node 1 to node 4 has been 'joggled' to save space. This does not affect the relationships and is often necessary.

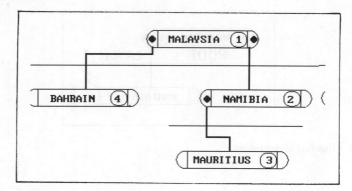

Figure 10.13   Developing the tree ....

10.16.11   Hong Kong is a *right*-descendant of Bahrain; Pakistan is a right-descendant of Namibia, whilst *nothing* in the remainder of the list comes 'before' Bahrain so *it* has no left-descendant. This process can now be continued until level 3 is completed as in Figure 10.14.

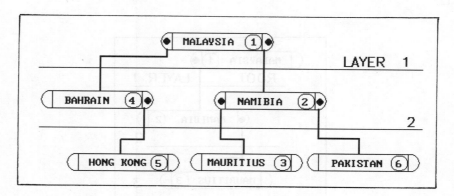

Figure 10.14   ... to complete layer 3

10.16.12   Now it is a matter of following the same pattern of decisions until the tree is complete. Can you do so by identifying *and* placing the remaining nodes and setting the pointers to identify left- and right-descendants respectively?

10.16.13   When you come to Botswana, which would appear to be a right-descendant of Bahrain since it comes after the latter, that position is already occupied, quite properly, by Hong Kong. You will need to 'slip down' a layer and attach it as a *left-descender* of the latter, which places it *between* Bahrain and Hong Kong, as it should be. Your finished result should resemble Figure 10.15.

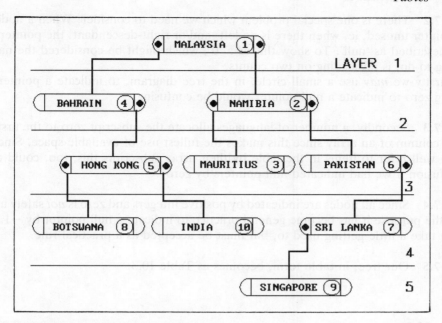

Figure 10.15 The tree diagram completed

10.16.14 There are a number of important elements here. One is the idea of 'layers' or generations. We see that 10 data are contained in a maximum of 5 layers in the present case and this means that no search, to find whether or not the name of a given country appears in the list, will take more than 5 'passes'. Can you see why? Try testing to see whether Dubai is in the list; how many questions or 'compares' do you need? How many to check whether Singapore is listed?

10.16.15 The second important idea is that each pointer directs us to a numbered node, ie to a *specific datum*. From the root every datum is on an identifiable sequence of branchings.

10.16.16 A third important consideration is that when a datum has *no pointers* we need search no further; we have reached a natural 'terminator'.

10.17 Tables

10.17.1 Drawing a tree is an excellent way of organising our data in the planning stage, with pencil and paper on the desk, but it is not at all suitable for storage and manipulation in the machine as it stands. The first step is to put *all* the information in the tree into a table, as seen below.

10.17.2   There is one special problem which we need to consider. When a node has a pointer unused, ie, when there is no left- and/or right-descendant, the pointer may be described as 'null'. To show this as a '0', which might be considered the natural thing to do, is misleading on two counts.

Firstly we *may* use a small circle, in the tree diagram, to indicate a pointer and using zero to indicate a null pointer would be confusing.

10.17.3   Secondly, a number of languages allocate the subscript zero to the first row and column of an array since this makes the fullest use of available space. Since we may well refer to items in the table by their array subscripts this, too, could cause confusion if we had indicated null pointers by zero.

10.17.4   Since all nodes are indicated by positive integers and zero is *not* safely usable for the purpose it has become general practice to indicate a null pointer by −1. This may take a little getting used to, but must be accepted as a practical rule.

10.17.5   Our tree, in table form, becomes as Table 10.3:

Table 10.3

NODE	DATUM	L. POINTER	R. POINTER
1	Malaysia	4	2
2	Namibia	3	6
3	Mauritius	−1	−1
4	Bahrain	−1	5
5	Hong Kong	8	10
6	Pakistan	−1	
7	Sri Lanka	9	−1
8	Botswana	−1	−1
9	Singapore	−1	−1
10	India	−1	−1

and we may now consider how to refine this into an acceptable application of our normal data structures.

10.17.6   At the system level, when used, for example, as a disk directory, the table, without the lines, can be stored as it is, so long as the length of each field is constant. This is the reason why many operating systems prescribe a *fixed length* file-name. Within a low-level language all symbols, whether alpha or numeric, are stored in numeric form and even the minus signs are inherent in the stored form as we saw in Chapter 6.

10.17.7   In most high-level languages however, since the data are STRINGS and the pointers are NUMERIC we have to seek other ways. The most obvious is to use an array since this already has a suitable rows and columns design.

However, we should need two MAPPED ARRAYS, one for strings, another for numerics, when modelling in this form a tree such as ours.

We *do* gain a little, however, since the ROW SUBSCRIPT will automatically identify the NODE, and the COLUMN SUBSCRIPT of the numeric array will tell us which are the left, and which the right, pointers. We can, therefore, both *save* one column and avoid the need for storing any labels to rows or columns.

10.17.8   As mapped arrays we shall have, for *our* tree:

$$
\begin{bmatrix}
\text{Malaysia} \\
\text{Namibia} \\
\text{Mauritius} \\
\text{Bahrain} \\
\text{Hong Kong} \\
\text{Pakistan} \\
\text{Sri Lanka} \\
\text{Botswana} \\
\text{Singapore} \\
\text{India}
\end{bmatrix}
\rightarrow
\begin{bmatrix}
4 & 2 \\
3 & 6 \\
-1 & -1 \\
-1 & 5 \\
8 & 10 \\
-1 & 7 \\
9 & -1 \\
-1 & -1 \\
-1 & -1 \\
-1 & -1
\end{bmatrix}
$$

in which all elements of our tree are reduced to the simplest possible model and are in a pair of *standard* data structures.

10.18   Back and Trace Pointers

10.18.1   We can go further and add BACK POINTERS to each node so that we may work *up* the tree from any node to trace its ancestors. As an example, node 10 of our COUNTRIES tree is descended from node 5 so that will be its back pointer. Node 6 has a *back pointer* of 2 since it is directly descended from that node.

10.18.2   We can also add a TRACE POINTER which, from any node, will allow us to trace the next in alphabetical (or numerical) sequence. Useful if, for example, we wish to list a directory in alphabetical order rather than in the alphabetically random sequence in which the files were created. In our example, Bahrain points to Botswana, since that is the next datum in alphabetical order; hence Bahrain's *trace pointer* will be

8. (We can either treat the trace pointer of Sri Lanka, the last datum in alphabetical order, as *null* and make it –1, or as taking us back to the head of the alphabetical sequence to indicate Bahrain with the trace pointer set to 4, giving us a cyclic pattern which we could enter at any node. The former is more common but the latter is often useful.)

10.18.3 Here we see the arrays completed in this fashion; can you see how it all fits?

$$
\begin{bmatrix}
\text{Malaysia} \\
\text{Namibia} \\
\text{Mauritius} \\
\text{Bahrain} \\
\text{Hong Kong} \\
\text{Pakistan} \\
\text{Sri Lanka} \\
\text{Botswana} \\
\text{Singapore} \\
\text{India}
\end{bmatrix}
\rightarrow
\begin{bmatrix}
4 & 2 & -1 & 3 \\
3 & 6 & 1 & 6 \\
-1 & -1 & 2 & 2 \\
-1 & 5 & 1 & 8 \\
8 & 10 & 4 & 10 \\
-1 & 7 & 2 & 9 \\
9 & -1 & 6 & -1 \\
-1 & -1 & 5 & 5 \\
-1 & -1 & 7 & 7 \\
-1 & -1 & 5 & 1
\end{bmatrix}
$$

10.18.4 Of the numeric array column 3 holds the back pointers and column 4 the trace.

As always familiarity will come only from experience and you should take a number of lists and try all the procedures for yourself.

10.18.5 Such a combination of arrays will not be operated on by any conventional matrix arithmetic, being merely data *holding* structures.

They are, in consequence, commonly referred to as TABLES and are of great importance at the system level. For example every disk has, as a fundamental part of its format, not only a directory but a File Allocation Table or Granule Allocation Table, showing how space is allocated to the file on disk.

There may also be a third part of the 'system information' written when the disk is formatted, the Hash Index Table, showing, from the hash code of the file name, exactly where the file is to be found in the directory. Even unusable tracks, detected on formatting, may be identified in a Track Lockout Table. (I have shown the capital letters which form, in each case, the acronym most commonly used, for example the FAT.)

10.18.6 These are all tables, of just the sort we have been investigating, with file names and various pointers to indicate disk cylinders, tracks and sectors, unallocated disk space and the like. With the utilities available to us today to rescue, read or amend files on disk an understanding of tables and how they work is indispensable, even for getting the best out of a PC.

10.19   Networks

10.19.1   The final topic to be looked at in this chapter is another which becomes more important by the year, networks. Just as some types of data fit 'naturally' into tree structures, so do networks have special characteristics.

10.19.2   Each is a structure which is marked by nodes and paths but a network is rarely so simple as to have a single root. The more common situation is to have several, or many, nodes with a variety of interconnections. As an example let us look at some simple computer networks.

10.19.3   Figure 10.16 shows four locations at which a company originally had computers installed on a stand-alone basis.

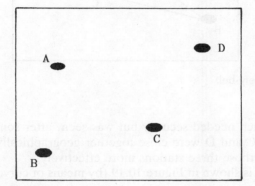

Figure 10.16   Four separate locations

When it was realised that much of the software used, the data-flows, the information distribution and the file-handling were common to all users it was decided to install a simple datalink between head office and each of the subsidiaries, Figure 10.17.

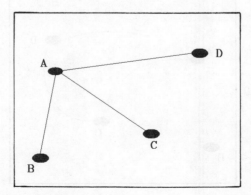

Figure 10.17   H.Q. to sub-stations

This worked well until, on several occasions, the system at A went down and B, C and D were left with the payroll not done on time.

The computer services manager then decided to duplicate some systems at B and to network B, C and D by additional links, shown in heavy lines in Figure 10.18.

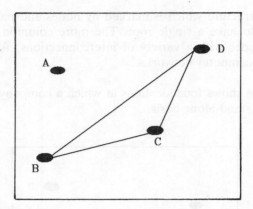

Figure 10.18   The safety-belt

This provided much needed security but was seen, after some time, to be rather slow and, since A, C and D were close together geographically, a high-speed LAN was set up to cover those three stations more effectively.

The new network is shown in Figure 10.19 (by means of curved lines to distinguish it from the others).

This too brought great improvements but to such an extent that the original slow link between A and B became an embarrassment and traffic around the other networks was too heavy to offer any relief to B.

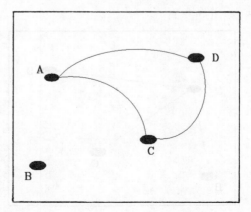

Figure 10.19   The Lan for A, C & D

By this time, however, technology had really leaped ahead and B and A, a considerable distance apart, were connected by a satellite link to extract maximum performance from all the systems as indicated in Figure 10.20.

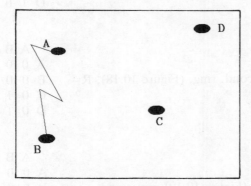

Figure 10.20   The final link

The final picture is shown in Figure 10.21.

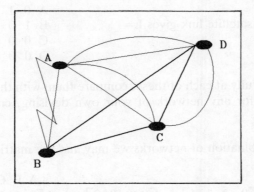

Figure 10.21   All the networks

10.20   Networks as Arrays

10.20.1   How could we model these different situations and evaluate the networks or cost out the solutions?

10.20.2   Once again the answer lies in using arrays of a rather special kind. The basic model is dominated by the fact that any single node may have a connection with any other node. What we need is a LOGICAL ARRAY using a 1 to represent 'connected' and a '0' for 'not-connected'. We can then model each network as a simple matrix.

10.20.3   For the first, star network, S=

```
      A B C D
    A 0 1 1 1
    B 1 0 0 0
    C 1 0 0 0
    D 1 0 0 0
```

For the second, ring, (Figure 10.18), R=

```
      A B C D
    A 0 0 0 0
    B 0 0 1 1
    C 0 1 0 1
    D 0 1 1 0
```

The LAN, (Figure 10.19), gives us L=

```
      A B C D
    A 0 0 1 1
    B 0 0 0 0
    C 1 0 0 1
    D 1 0 1 0
```

Lastly the satellite link gives L=

```
      A B C D
    A 0 1 0 0
    B 1 0 0 0
    C 0 0 0 0
    D 0 0 0 0
```

10.20.4   Look carefully at each of these, compare them with the diagrams and make sure that you can, for any network of your own devising, create the appropriate matrix.

10.20.5   For a combination of networks we may *add* the matrices.

$$
\text{S+R, for example:} \quad
\begin{bmatrix} 0 & 1 & 1 & 1 \\ 1 & 0 & 0 & 0 \\ 1 & 0 & 0 & 0 \\ 1 & 0 & 0 & 0 \end{bmatrix}
+
\begin{bmatrix} 0 & 0 & 0 & 0 \\ 0 & 0 & 1 & 1 \\ 0 & 1 & 0 & 1 \\ 0 & 1 & 1 & 0 \end{bmatrix}
=
\begin{array}{c} \\ A \\ B \\ C \\ D \end{array}
\begin{array}{c} A\ B\ C\ D \\ \begin{bmatrix} 0 & 1 & 1 & 1 \\ 1 & 0 & 1 & 1 \\ 1 & 1 & 0 & 1 \\ 1 & 1 & 1 & 0 \end{bmatrix} \end{array}
$$

10.20.6   In this case the matrix may be seen as wholly logical, since each 1 indicates that there *is* a path from one node to another. It is, however, also an *arithmetic* matrix since the 1 also says that there is *only one* such path.

What happens if we now also combine network 'L'?

On the basis of the *arithmetic* we now see that there are, for example, *two* paths from A to C. (In these examples all paths are assumed to be two-way channels. Can you see how to indicate one-way paths?)

10.20.7 We can, in fact, use network matrices *either as logical or as arithmetic* tools as long as we don't confuse the two very different types of process.

10.20.8 Can you complete the S+R+L matrix as well as all those representing any combination of two or more of our example networks? What is the array which represents all the networks together, as in Figure 10.21?

10.20.9 As a final exploration of the ideas examined in this chapter, discuss with your colleagues how to exploit these models. Invent and model some networks of your own. Prepare simulations of the traffic on each network, inserting numbers of your own choosing to identify the 'data-packets-per-day' flow rates. Allocate a price per packet for each path on each network and *cost* the traffic. Identify any intrinsically weak or possible redundant links in the system.

10.21 Conclusion

Throughout this chapter we have seen how common *existing* mathematical techniques provide ways and means of creating and manipulating data structures. We have also seen the need to practise working with them to achieve the understanding and fluency of their use which will allow us, from the earliest design stage of programs, to *select* existing and reliable ways of doing things rather than having to 're-invent the wheel'.

Exercises

1) A manufacturer produces three different machines, A, B and C each of which embodies different arrangements of the sub-assemblies, 'd' and 'e'. Each of these, in turn, is made up of different quantities of parts having the part numbers 101A, 101B and 101C and which cost $3, $2 and $4, respectively.

   Model A consists of 3 of sub-assembly 'd'
   and 3 of sub-assembly 'e';
   Model B consists of 2 of 'd' and 4 of 'e';
   Model C has 3 of 'd' and 5 of 'e'.

   Of the sub-assemblies:
   a 'd' needs 3 of 101A, 4 of 101B and 3 of 101C
   an 'e' needs 2 of 101A, 3 of 101B and 5 of 101C.

   a) Put all these data into appropriate matrices.

   b) What is the matrix of costs/machine models?

   c) i) In one day, 80 of A, 50 of B and 40 of C can be produced. Set these data into the appropriate vector.

      ii) Find, by matrix operations, the total value of one day's production.

      iii) If the costs of the parts 101A, 101A and 101C were increased by 5%, 7½% and 4% respectively what would be the total increase on one day's production costs?

2) Why are linked lists often used instead of arrays?

3) Why are such lists linked?

4) i) What is the function of a pointer?
   ii) Show, in a diagram, a linked list of your own devising complete with pointers.

5) In what ways do LIFO and FIFO stacks differ?

6) i) Identify the characteristics of one kind of sort.
   ii) Illustrate your chosen example by a suitable diagram.

7) Illustrate the process by which a binary chop search would find item 77 in an ordered file of 150 records.

8) a) Construct a data tree to hold the data: 497, 235, 502, 50, 79, 143, 64, 379

   b) Create, for the same tree, a table with the appropriate right- and left-pointers.

   c) Add the necessary back- and trace-pointers.

   d) Show how the table in part c) would look after the addition to the tree of 128.

9) The following matrices, A and B, show communication links between three separate locations.

$$A = \begin{bmatrix} 1 & 0 & 0 \\ 1 & 0 & 1 \\ 0 & 2 & 0 \end{bmatrix} \qquad B = \begin{bmatrix} 0 & 1 & 0 \\ 1 & 0 & 1 \\ 1 & 0 & 1 \end{bmatrix}$$

   a) Draw the corresponding networks.

   b) Add the matrices and draw the resulting network.

   c) Which, if any, of the communication links are one-way only?

10)
$$\begin{bmatrix} 0 & 1 & 0 & 1 \\ 0 & 0 & 1 & 1 \\ 0 & 0 & 0 & 1 \\ 1 & 1 & 1 & 0 \end{bmatrix}$$

is the network which at present connects various workstations to a central file-server.

   a) Label the matrix and show which is the central computer.

   b) It is planned that the three 'outstations' shall have full local processing capability but shall become fully linked to each other by two-way communications whilst remaining connected to the central file-server.

   i) What must the new matrix be?

   ii) Write down the matrix which shows what new paths are needed.

11

Numerical and Algebraic Methods

Objectives

After working through this chapter you should be able to:
- use elementary numerical and algebraic methods to model and simplify problem solving;
- understand iterative, recursive and convergent processes.

11.1 The Background

11.1.1 I have said earlier in this book that the mathematicians of old dreamed of having a calculating device of real power. Among their number were people such as Babbage in England, Pascal in France, Leibnitz in Germany and, more recently, Hollerith in the United States.

11.1.2 In the absence, until very recently, of any truly successful device, much important work in modelling the real world and solving real problems was both laborious and time-consuming.

11.1.3 Mistakes in calculation were also a major difficulty since with paper and pencil, or even small mechanical calculators, the human being was, and still is, a relatively unreliable part of the chain.

11.1.4 Throughout many centuries the ultimate mathematical skill was to reduce a problem to an *algebraic model*, or a Euclidian *geometric model* and to operate on these models in essentially *abstract* ways.

11.1.5 These were rarely of use to the builder or the engineer and there were many such models which could not be fully 'solved' or proved, or which could not be reduced to a suitable form for real-life applications.

11.1.6 There were also many problems to which no *exact solution* could be found but 'final' solutions, full proofs or exact solutions are not necessarily what any builder, engineer or accountant needs. We need, for most practical applications, *working values* which allow the persons concerned to get on with solving their real-life problems.

11.1.7   It is in meeting this need, in *real* applications of mathematics, that the roots of the development of *numerical* methods lie. Today, in many applications of the digital computer, they come into their full usefulness.

11.2   Language

11.2.1   There are, as always, some special words and symbols associated with the topic and among these are ITERATIVE, RECURSIVE, CONVERGENCE and LIMIT as well as, from Chapter 5, ERROR and ERROR BOUND.

11.2.2   We will deal with each of them more fully as we meet them in action but 'iterative' really only means 'repetitive' and an 'iteration' is a process which repeats. Most commonly it repeats the same core calculations to arrive at ever-closer approximations to the value being sought.

11.2.3   We should not fall into the common trap of assuming that an 'approximation' is in any way an inferior solution to a problem. It certainly does not mean a wild guess, or even an estimate. An approximation is often the only useful or usable solution which exists! For pi, for example, we can never have a final value, only an approximation to a sufficient absolute error bound to satisfy the practical needs of the user.

11.2.4   Every rounding, truncation or 'significant figures' answer results in an approximation. *The key is the level of accuracy which is of practical value to the user.* For a great many calculations of all kinds we define the appropriate degree of refinement of approximation by defining the acceptable error bound. A size calculated correct to 7D, as in seven figure tables, for example, is of very little use to an engineer whose instruments or machines have an inherent error bound of ± 0.001mm. He may, however, need such precision *within* calculations to minimise error.

11.2.5   'Recursive' for our present purpose means only that the product of any step of an iteration is 'fed back in' to the calculation for the succeeding step. When successive values from an iteration show *decreasing differences*, we say that the iteration is 'converging' and it is not difficult then to interpret '*di*vergence'.

11.2.6   We met the word 'limit' earlier when it described the theoretical end of a process, such as finding the value for 'pi', even though we may never be able to arrive at a finite answer.

11.2.7   We may also limit the number of repetitions of an iteration by defining the degree of accuracy, to which it should pursue its calculation, by identifying the acceptable error.

11.2.8   The principle new *symbol* which we shall use is Δ, from the Greek letter 'Delta', representing the *difference* between any two numbers or successive terms in a sequence.

11.3 Iteration

11.3.1 Let us examine a simple iterative process and see what happens, in a pencil-and-paper 'run' and tabulating the results of each 'pass' round the loop. These are the vital techniques which should be followed when exploring any iteration:

$$
\begin{aligned}
&i=1,\ d=1,\ s_1=1 \\
&\text{ACCEPT } n \\
&\qquad \text{DOWHILE } i \leqslant n \\
&\qquad\qquad \text{Print } i,\ s_i \\
&\qquad\qquad i \leftarrow +1 \\
&\qquad\qquad d \leftarrow d+2 \\
&\qquad\qquad s \leftarrow s_i - d \\
&\qquad \text{ENDDO} \\
&\text{END}
\end{aligned}
$$

11.3.2 Table of results

Table 11.1

PASSi	d	Print i	Print s_i
1	1	1	1
2	3	2	4
3	5	3	9
4	7	4	16
5	9	5	25
6	11	6	36

11.3.3 This table contains some repetitions since 'i' is repeated in the print-out but it is the *pattern* of table which we should follow. It will, in many cases, be necessary to do more than six 'sample' passes in a run but six is enough for us here. What does the 'print' information tell us? What function maps the 'i' onto the 's_i' ?

11.3.4 Clearly it is the table of natural squares which we are developing so how did I know that the model in the pseudo-code would generate that function? Let us take only the 'print' columns and add a differences (Δ) column so that we may fully explore the iteration. (The table of any such sequence should, ideally, start at 0 since this is the origin of all our natural numbers.)

The table including differences:

Table 11.2

i	0	1	2	3	4	5	6
s_i	0	1	4	9	16	25	36
$\triangle$	0	1	3	5	7	9	11

and we see that the sequence of differences is the set of odd numbers. In other words, the natural squares of numbers up to 'n' may be found by adding all the *odd* numbers from 1st to nth. It is not difficult to see how the table would continue.

11.3.5   There are many ways of modelling this situation and one of the traditional mathematical models is some variant of:

$$n^2 = \frac{n}{2} * (1 + (2n - 1))$$

11.3.6   This is a fair comparison between 'classical' mathematical solutions and the straightforward, direct, iterative model in our pseudo-code segment above. Such a model is usually both the simpler and the more efficient way of using the power of the computer and does not involve memorising formulae.

11.3.7   Note that in this case we are dealing with the natural squares so there is no question of approximations in our iteration.

11.4   An Iterative Model

11.4.1   There is not a set of mathematical tables stored in the computer to provide 'sqrt(7)', for example. How could we model a process for finding $\sqrt{7}$? Let us try a few 'what if' steps and see what happens.

11.4.2   What if we take a first estimate, e_1, of $\sqrt{7}$ as being 3.5, simply by halving 7? If we now divide 7 by e_1 the answer will identify another factor of 7. If the *second factor* were to turn out also to be 3.5 then that would indeed be $\sqrt{7}$. Instead it is 2.

11.4.3   We can now see that 3.5   is a larger factor of 7 than is $\sqrt{7}$ and 2 is smaller so $\sqrt{7}$ must lie somewhere between 2 and 3.5. We can try the simple mean of these two values, 2.75 as a closer approximation. Now let us turn to the calculator.

11.4.4 If we now divide 7 by our improved estimate, 2.75, we get 2.5454545, and again we know that the true root must lie between these two values. Once more we 'average' these, giving 2.6477272 and 7 divided by this gives 2.6457520. Averaging these two gives 2.6457506; the next repeat gives 2.6457513 and dividing 7 by this gives 2.6457513. We have, in a very few steps, with a calculator giving seven places of decimals, found $\sqrt{7}$ correct to 6D.

11.4.5 If e_i is each successive estimate and 'n' is the number whose root we seek, then $e_1 = n/2$ and, putting the steps above into a formal model:

$$e_{i+1} = \frac{(e_i + (n/e_i))}{2}$$

we have the core of our program segment.

11.4.6 Test this model carefully, using any, or all, of the steps detailed above, and with 7 as 'n'.

11.4.7 Now to tabulate those steps, *with* the differences column:

Table 11.3

i =	e_i	n/e_i	e_{i+1}	$e_i \triangle e_{i+1}$
1	3.5	2	2.75	0.75
2	2.75	2.5454545	2.6477272	0.1022728
3	2.6477272	2.6437768	2.6457520	0.0019752
4	2.6457520	2.6457506	2.6457513	0.0000007
5	2.6457513	2.6457513	2.6457513	0.0000000
6				

11.4.8 If we now plot on a graph the successive values in columns 2 and 3 of our table, of the estimated root and its 2nd factor of 7, we can see the rapid convergence of such an iteration, as suggested by the rapid decrease in the differences, in that column of our table. See Figure 11.1.

11.4.9 This is the typical behaviour of such a routine and it is often extremely valuable when programming to arrive, initially, at such a very simple model, sometimes referred to as a 'boiler-plate' version, which is simple, works effectively and is easy to test.

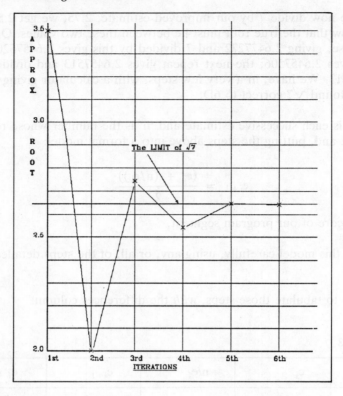

Figure 11.1 The convergence is clear

11.4.10 The model is fast, even if our initial estimate is, as in our example, very crude. An initial estimate, sometimes referred to as a 'seed', *must* be provided by some means and many textbooks recommend putting in an estimate 'by hand', so to speak. It is, however, extremely easy to halve our 'n' in order to arrive at a first estimate and it is often the case that a very simple method is good enough.

11.4.11 Once we have established the working core of our program we can, if need be, refine our model or add other steps to it, (for example to read in *any* 'n'). We may even seek out a more elegant or more efficient version from classical mathematics texts. Any specialist textbook on Numerical Methods will contain a wealth of such models, so long as we know just what we are looking for.

11.5   Limiting the Iteration

11.5.1 We need to add a 'stop' to our procedure, and some tests, as it is all too common for a computer to 'lock up' when performing an iteration which has no natural end or which is, in fact, *diverging*. The 'infinite loop' is a long-standing hazard in programming!

11.5.2 The hazard of divergence is best dealt with by comparing successive *differences*. As we see clearly in our *convergent* example the differences *always* decrease. If, on any pass, the difference has *increased* the iteration is diverging, and if this is unexpected we must cause our program to break out of the loop and let us know what is happening.

11.5.3 In order to stop the procedure when it has gone far enough for our purpose, we must first define, with the user, what that purpose is and *to what level of accuracy of estimate* the answer is to be calculated.

11.5.4 If we need an answer correct to 3D, what will be the absolute error bound? If you are in doubt refer back to Chapter 5.

In order to be sure that we are rounding correctly we would ordinarily calculate to 4D (but, for some purposes, we may wish to calculate to as much as, but certainly no more than, 5D).

11.5.5 We achieve this by controlling our loop with a statement such as:

$$\text{DOWHILE abs}(e_i - e_{i+1}) > .0001.$$

11.5.6 You may now feel that you could write a pseudo-code routine for 'sqrt(n)' which embodies a test for divergence and the 'stop' when an answer can be given correct to 3D.

Try it; vary the 'n' and test each version first by a desk-top run using your pocket calculator and then as a short program in your chosen high-level language on a computer.

As always, mathematicians have devised faster methods for finding roots and further study will lead you to these if you are interested.

11.6 Aberrations

11.6.1 The problem of controlling an iteration should not be underestimated and much of modern 'chaos theory' has evolved from the study of iterative routines which produced unexpected results.

Often, such aberrations show up as an unexpected failure to converge or a lapse, at an unexpected point, into results of disorderliness where order was expected.

11.6.2 We must, therefore, be meticulous in auditing the operation of an iterative process. We can do this by printing full listings of intermediate results if necessary; by graphing sample sets of results and looking for anomalies in the graph; by continuing the iterations longer than may seem strictly necessary; by paying careful attention to tables of differences or by trying 'unorthodox' seed values or step sizes.

11.6.3   Test data should always include unlikely or even impossible values to make sure that they are detected by our programs before generating misleading reports.

11.7   Simultaneous Equations

11.7.1   Let us look at a rather more complex type of problem and see what we can do with it.

11.7.2   Here is a pair of simultaneous equations:

i) $2x + 3 = 5$
ii) $x^2 + y^2 = 4$

for which we are to find *positive* values of 'x' and 'y' which satisfy both. (Since we want only positive values I have shown only the first quadrant of the graph.)

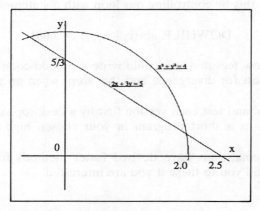

Figure 11.2   A pair of simultaneous equations

11.7.3   You may find it useful to draw the graph to a large scale for reference as we work through the example.

11.7.4   This is exactly the kind of problem which presents quite difficult algebra but yields to an heuristic, iterative, numerical approach. The (+,+) point at which the straight line cuts the circle is one at which both equations are true.

11.7.5   At that point 'x' is clearly a little less than 2. If we take a 'seed' value of 1.5 and substitute this for 'x' in equation i) we shall find that $y=0.66667$ to 5D. Substituting this for 'y' in equation ii) we find that $x=1.88561$ and we take this as x_2 for our second pass and substitute it for x in equation i) to find y_2, which, in turn, we substitute in equation ii) to find x_3, and so on. NOTE that, in all these cases, we must be careful not to confuse the *subscripted* values of x or y with *powers* of those values.

11.7.6 In Table 11.4, I have set out the sequence of values at each pass and you will see that, even for a computation of admitted algebraic complexity, the numerical method leads us to a result very quickly indeed, and we see that $y = 0.3544$ and $x = 1.9683$ satisfy both equations.

Table 11.4

i	Eq'n i) x	Eq'n i) y	Eq'n ii) y^2	Eq'n ii) x
1	1.5	0.66667	0.44449	1.88561
2	1.88561	0.40960	0.16777	1.95761
3	1.95761	0.54239	0.29419	1.92505
4	1.92505	0.38330	0.14692	1.96293
5	1.96293	0.35804	0.12820	1.96769
6	1.96769	0.35487	0.12594	1.96826
7	1.96826	0.35449	0.12566	1.96833
8	1.96833	0.35445	0.12563	1.96834
9	1.96834	0.35444	0.12563	1.96834

11.7.7 You should now try to write a program segment to include this routine based on a simple loop construct. The model which we have been looking at is clearly both recursive and convergent, and this should guide us in our programming.

11.7.8 Now try putting into practice an alternative version of the model, we could call it 'routine 2', in which you put the seed value into equation ii) and transfer the result to equation i) each time. What do you notice about the results?
If you arrive at $x = -0.4299$ and $y = 1.9532$ to 4D, we agree.

11.7.9 Look again at the graph; can you see what these values might correspond with?
They are the x and y values which satisfy the two equations at the other intersection.

11.7.10 However, we *specified* at the outset that only the +ve values were applicable, so that if we had tried only this second method we would have defective answers.

11.7.11 We could avoid this by including in our model a standard test couched in terms something like:

'If x $\leq$ 0 OR y $\leq$ 0 then CONTINUE else DO routine 2'.

As mentioned earlier, the auditing of our model is very important and will often cause us to add several conditional tests to a simple routine.

11.8   Expansions

11.8.1 Many functions are EXPANSIONS which seem a little different at first but which are, again, not too difficult to program. For example a very useful approximation to π is to be found by the model:

$$\pi \approx 4 * \left(\frac{1}{1} - \frac{1}{3} + \frac{1}{5} - \frac{1}{7} + \frac{1}{9} - \frac{1}{11} \right)$$

which is really quite easy to write, since we can predict further terms without difficulty, but rather tedious to work by pencil-and-paper arithmetic.

11.8.2 Left to itself, such an iteration will recur indefinitely and so we need, once again, to consider the level of precision which our application needs in order to set a 'break-out' difference between successive terms, which will stop the iteration when it has gone far enough. If you try to model this, you may be surprised at the ease of calculating π, although the iteration is quite slow compared with our $\sqrt{n}$ model, above. However, *absolute* speed does not really concern us since a thousand passes round such a loop (and all iterations involve the loop construct) take little perceptible time when it is the computer which is doing the 'number-crunching'.

11.8.3 Virtually all the angle functions, logarithms and many other mathematical 'tables' are calculated using such iterative expansions, although, as with π, mathematicians have provided us with faster and more powerful models than some of the relatively simple ones at which we have looked in this chapter.

11.9   Curved Areas

11.9.1 The last topic in this section of our work deals briefly with some questions concerning curved-line graphs. There are two questions which arise frequently when we need to deal with functions of the kind which we saw, in Chapter 7, to generate curves when plotted on a graph.

11.9.2 The first such question is, "How may we INTERPOLATE or EXTRAPO-LATE from what we know?"

11.9.3 The second question is, "What is the area of a surface, like a graph, which has one curved edge?"

11.9.4 In the past, both such questions were dealt with by rather elaborate methods but we may, for most purposes, use our iterative processes to good effect.

11.9.5 If we take, as a basis for experiment, the graph of $y = ax^2 + c$ we can write a standard program module to calculate the curve:

```
Read x, breakout, step, a, c
      DOUNTIL x=breakout
          y₁ ← (a*(x ↑ 2))+c
          print x, y₁
          x ← x+step
          y₀ ← y₁
      ENDDO
   END
```

We can now calculate the points of our graph by initialising the variable 'x' to our first value, 'breakout' to the last required value and 'a' and 'c', which are constants for any given function, to their fixed values.

11.9.6 Furthermore, we may make our curve as smooth as we need by using a very small 'step' value, that is the amount we add to our previous value of 'x' to derive the next value.

11.9.7 As a little 'bonus', if, as we stated earlier, we wish to find the AREA under the curve between any two values of 'x' the step size will also be the *width* to be used in our calculation. Let us look at the diagram of a small section of the area under a curve in Figure 11.3.

11.9.8 To find the area, 'A', under the curve between any two 'x' values, using 'firstvalue' and 'secondvalue' as the variable names for the two values of 'x', for the moment, we can insert into our program segment a sub-routine of the type:

```
DOWHILE x ≥ firstvalue AND x ≤ secondvalue
     A ← A + step*½(yᵢ + yᵢ₋₁)
ENDDO
```

11.9.9 We can do this because, as we see from the diagram, the area between, say x_7 and x_n, (lightly shaded in Figure 11.3), may be seen as the sum of a number of strips, of which a specimen is shown with diagonal hatching in the diagram.

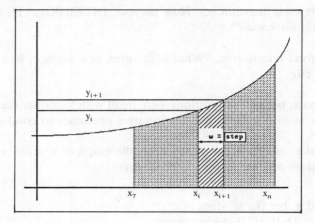

Figure 11.3   The area as strips

11.9.10   Each strip is a close approximation of a TRAPEZIUM for which the AREA is the width multiplied by the *average* of the two parallel sides.

11.9.11   In our example, our main routine gives us sequential pairs of the '*y*' values, the heights, and the step size gives us the width.

   Since all three occur in each pass of our original loop, it is only a matter of 'capturing' existing data and accumulating partial areas under 'A'.

11.10   Interpolation

11.10.1   If we wish to *interpolate* further values between values for '*y*' which we have already plotted, or *extrapolate* from them we need only set up our program segment so that we may *enter specific values of* '*x*' from the keyboard.

11.10.2   As always, it is practice with such routines which will bring confidence and, above all, don't be afraid to try out variations, such as finding '*x*' from an input value for '*y*'.

11.11   Some Algebraic Methods

11.11.1   We have found ourselves increasingly using *symbols*, rather than numbers, and constructing *complete* rather than partial models. It seems that we have reached a natural point at which to look at some ALGEBRAIC METHODS.

11.11.2   Let us first of all remind ourselves that the main purpose of algebra for us, as for most *users* of its techniques, is to allow us to create COMPLETE MODELS, which we can then program.

11.11.3   Such models will include all *known* information whether CONSTANTS or VARIABLES, which will eventually allow us to isolate and determine the *unknown(s)* which represent the usable solution to a problem.

11.11.4   In computing, it is usually worth developing such a model only if it represents a problem which arises frequently, such as calculating employees' pay. Not just *one* person's pay, but any number of people's on any number of occasions.

It is *this* which leads us to algebra and its methods. If *my own* pay were $10 per hour and I had worked 40 hours with $15 taxes to be deducted the *arithmetic* model, $(40*10)–15 would be sufficient.

11.11.5   In order to GENERALISE the model, that is to say, make it useful for *any* worker in *any* week for *any* tax deductions, we would have to describe a more general model such as:

Rate times Hours minus Taxes is the take-home Pay.

This is a true representation of the situation but still rather cumbersome and easy to mis-read, so we reduce it to *abstract symbolic form*, and say $R*H - T = P$.

This is a true ALGEBRAIC MODEL; a FULL statement of the model in SYMBOLIC FORM.

11.12   What the Model Means

11.12.1   In one cell of a spreadsheet which I use frequently appears a 'formula', as it is described in the language of spreadsheets, which seems quite complex:

@INT(B1 + 0.5   + (F1 – A1) * (D1 / C1)).

We should not ordinarily recognise this as an equation since it lacks the very symbol which says 'equation', namely the '=' sign to denote that the quantity to its left has precisely the same value as that to its right. We have seen earlier that there are also many 'inequations', or 'inequalities' denoted by '$<, \leq, >, \geq, \neq$' and also many approximations denoted by '$\approx$' (or $\simeq \doteqdot$ perhaps, depending on the country) and we must by no means forget these.

11.12.2   In our spreadsheet example the equivalence is IMPLIED, that is to say that, although it is not *stated*, the cell in which the formula is placed is, effectively, the left-hand side of the equation and the destination for the value to be calculated.

11.12.3   What all this amounts to is that there is a specially adapted *spreadsheet algebra*.

11.12.4   Let us look at it in detail and find out what all the parts of
@INT(B1 + 0.5   + (F1 − A1) * (D1 / C1)) actually mean.

11.12.5   First we can identify the @ symbol as being one of the ways in which spreadsheet software identifies a CALCULATION WHICH IS TO BE PERFORMED with the result being placed in the cell where the formula is located.

11.12.6   'INT' is a common computer-language command to take the INTEGER PART only of the answer by *truncating* at the decimal point.
  Hence INT(3.845618) would return the value '3'.

11.12.7   The rest of the model is in brackets and so must be dealt with by the 'rules of brackets and the sequence of operations'. These state that the contents of brackets are to be dealt with in sequence from the *innermost to the outermost* and that arithmetic operations are to be done in the sequence:

  1) Raise to power, (x^2), or $\sqrt{}$;
  2) * or /;
  3) + or −

  and when two operators are of equal 'precedence', say 12*4/3 they are to be performed in sequence from LEFT to RIGHT.

11.12.8   If we are in any doubt about precedence, especially when designing a program model, the remedy is to break the model into single operations within yet more pairs of brackets. Even many pocket calculators today will allow us to 'nest' brackets 6 'deep'. One precaution is to make a final check to ensure that the number of left-brackets '(' exactly equals the number of right-brackets ')'. Many seemingly mysterious error reports arise in computing because of a failure to check this.

11.12.9   We can label the model to indicate the steps in which we first solve (F1−A1); next (D1/C1); then what is inside the outermost pair of brackets, shown with the 'long' underline until *finally* we can take the INTEGER value of the result. This is shown in Figure 11.4.

$$\text{INT}\ \frac{(B1\ +\ 0.5\ +\ \overset{\text{1st.}}{\overline{(F1\ -\ A1)}}\ *\ \overset{\text{2nd.}}{\overline{(D1\ /\ C1)}})}{\text{3rd}}$$

Figure 11.4

11.12.10 There remains the mystery of the F1, A1, D1 and so on. In the algebra of spreadsheets these are the CELL ADDRESSES in which the variables are stored and to this extent they are precisely similar to the storage of *declared variables or constants* at fixed addresses in the computer's memory. They are, then, just VARIABLE NAMES rather than the '*x*', '*y*' and '*z*' of conventional, wholly abstract, algebra.

11.12.11 We can now see that (F1–A1) says "Take the value stored in A1 and subtract it from the value stored in F1." What is *not* said is that the result will have to be put into *temporary storage* until it can be used again later in the calculation. Such temporary storage, exactly like writing intermediate values on a piece of paper in everyday practice, is often referred to as SCRATCHPAD memory and many applications of our own programs need some form of scratchpad, or temporary, 'holding' variables, built into our planning.

11.12.12 In the spreadsheet all the manipulations of such models are transparent to us but it is *we who have to design the models*.

11.12.13 It is exactly here that we meet the need to have some understanding of algebraic methods; they are, quite simply, the language of modelling for the computer. The original of the spreadsheet operation which we have been considering is the equation:

$$x'_i = \bar{X}' + (x_i - \bar{X}^*(\sigma'/\sigma))$$

11.12.14 We may now be able to see how this original 'maps onto' the spreadsheet version. Discuss this with your colleagues, try some values for all the *x*'s and σ's and see also if you can work out what the 0.5 is for in the spreadsheet version.

11.13 Notation

11.13.1 When looking at any symbolic model the most important thing is not to be put off by the symbols themselves. They all merely *stand for* something and are not sacred objects! What we need is patiently to unravel these labels and, if they are unfamiliar to us, to look them up in a reference book.

The symbol 'σ', for example, is commonly used in statistics for Standard Deviation but we could, for our own purposes, use any symbol we choose. The great advantage of using 'σ' is that it is understood anywhere in the world as S.D. when it is used in a statistical context.

11.13.2 In a private model or program we could perfectly well call it 'Fred' if we wished! What we MUST do, for the sake of *maintainability* of our programs, is to include our chosen symbol, the word or words which it stands for, the type of variable used and any other relevant information in our DATA DICTIONARY. Even we, looking at a program which we wrote a few months earlier, may find it difficult to interpret unusual symbols or 'Freds' if we did not document them properly.

11.13.3   In traditional mathematics the documentation takes the form :

'Let x = the rate of pay; let y = the number of hours and let z = the amount of tax'.

11.13.4   The use of x, y and z is a tradition in formal algebra largely because the early letters of the Arabic alphabet were already being used for arithmetic or geometric 'identities' such as A for area, B for breadth, C for circumference, and so on.

Since letters up to 't' for time were already in use for *known* values, mathematicians adopted the practice of using x, y and z for the *unknowns*. For several centuries, three-variable relationships were the most complex to be tackled and these three symbols served, and continue to serve, well enough. You must not let these old habits worry you if algebra is not one of your favourite pastimes; any symbol which is appropriate to the task will do equally well.

11.13.5   We have only two other basic ideas of notation to discuss at this stage and the first is the convention for multiplication. There are *two* common ways of indicating multiplication; $a \times b$ and $a * b$ but we have already seen that the first, older version, is of no use in computing since we have no means of distinguishing between the *quantity* 'x' and the *operator* 'x'. In mathematics if something *can* be misunderstood it is treated as *wrong* since, sooner or later, it *will* be mis-read or mis-interpreted.

When 'x' came into use in algebra a similar problem arose and multiplication was indicated by '$x.y$', as it often still is today.

Little by little, especially with the invention of decimal fraction notation, it became clear that even the dot could be ambiguous and it was not safe to write 3.5, for example, to mean 3 *times* 5.

11.13.6   The solution adopted was to establish that where different symbols represented quantities to be multiplied they could be written side-by-side without any intervening operator. Nowadays it is common practice to write x*y simply as *xy* but *not all translator software recognises this* so be cautious!

11.13.7   In similar fashion $x(y+z)$ means that *all the contents* of the brackets are to be multiplied by 'x' either before enumeration, as $xy + xz$, or, if we know the values to 'substitute' for x, y and z, say 2, 7 and 5 respectively, $2(7 + 5)$ we may work as $2 \times 7 + 2 \times 5$ *or* $2(12)$, whichever is the more convenient.

11.14   Powers of Numbers

11.14.1   We then have a very confused situation if we wish to multiply a number by itself, for 'n' times, and we try to write it as, say, xxx. Do we have x*x or x*x*x? If x were 5, for example, the answer could be 25 or 125 depending upon our interpretation.

11.14.2   This clearly will not do. So we borrow from the very old ideas of 'powers' of numbers in our column system and in square numbers and their relatives and use x^2 to denote x*x, $3x$ to denote x+x+x and x^3 for the x*x*x.

11.14.3   This INDEX NOTATION, or 'powers of n' is very common in algebraic models and leads to a number of special laws governing the use of 'indices', which is the plural of 'index'. In the notation x^n the small numeral in the 'top-right-corner' of the term 'x' is its INDEX and indicate how many x's are to be multiplied together.

11.14.4   The first new law is that 'when like terms in index notation are multiplied together their indices are *added*.' For example $x^2 * x^3$ is equivalent to x*x * x*x*x or x*x*x*x*x which is x^5.

11.14.5   Here I have mentioned, for the first time, the phrase 'like terms', which is important in algebra.

In, for example, $2x^3 + x^2 + x + 5x^3 + 3x + 4x^2$, we see that all the terms are similar since they all include 'x'. However x^2 is not exactly comparable with x^3 since it is on a different scale.

11.14.6   To simplify the original pattern we can, however, rearrange the order so that we have $2x^3 + 5x^3 + x^2 + 4x^2 + x + 3x$.

We call this 'gathering like terms' which we do, both by tradition and for clarity, in the ordered pattern of alphabetical order, e,g, 'x' before 'y', and within that, descending order of indices.

11.14.7   It is not too difficult to see that we now have, grouping the terms altogether, $7x^3 + 5x^2 + 4x$, which is rather simpler than our original version.

11.14.8   Such SIMPLIFYING is at the very heart of algebraic methods. When we have a number of terms in which different symbols are multiplied together, for example $3abc + 2ab + 5cba$ it may be a little more difficult to decide which to 'gather'.

Only the '*abc*' and the '*cba*' (or any *other* arrangement of these three letters) are 'like' since c*b*a comes under the ASSOCIATIVE LAW and could equally well be written as c*a*b, or c*b*a.

Our gathering of like terms, then, will result in $8abc + 2ab$ but, by the DISTRIBU-TIVE LAW, since we have, in effect $(ab * 8c) + (2 * ab)$, we *could* write this as $ab(8c+2)$.

11.14.9   Think about this, discuss it with your colleagues, try a variety of such associations and distributions until you are confident with them. They are all ways in which we can simplify our models.

The laws of arithmetic operate just as powerfully and to great advantage in algebra. We shall make use of them very frequently.

11.14.10   Returning to our work on index notation, it is the laws of INVERSES which have special importance. The additive inverse of 2 is –2 and x^2 has the additive inverse x^{-2}, which represents $1/x^2$. You can see this in our earlier work on the column values of our ordinary number system.

The multiplicative inverse of x^2 is $x^{1/2}$ which is the SQUARE ROOT of x. All this leads to some manipulations of indices which need a good deal of practice to develop real confidence. Take $x^2(x^3+2x^2+x+2)$ for example, which, when multiplied throughout will become $x^5+2x^4+x^3+2x^2$.

11.14.11 We do occasionally meet situations of this kind in modelling real-life problems and a working understanding of index notation is valuable.

11.15 Simplifying

11.15.1 Of much greater importance, however, is the ability to simplify our models and to change their form to suit our needs.

11.15.2 First the simplification process.

At the heart of any equivalence (equation) is the requirement that *we do nothing to disturb the balance* which that equivalence demands.

Both sides *must* remain of equal value whatever we do to them. Eventually we would like to see, simply, $x = n$.

11.15.3 The golden rule when simplifying is to remove one complication at a time by performing the *inverse operation* with the *inverse value* on *each side* of the equation.

11.15.4 Let us do so systematically, in steps, starting with a simple example:

$$\tfrac{1}{2}x + 14 = 19$$

Step 1 $\tfrac{1}{2}x + 14 - 14 = 19 - 14$

brings $\tfrac{1}{2}x = 5$

Step 2 $2*(\tfrac{1}{2}x) = 2*5$

brings $x = 10$

11.15.5 The process works for complicated models also:

	Look at	$7x^2 + 15x - 5 = 5x^2 + 3(5x+15).$
1	Remove the brackets:	$7x^2+15x-5 = 5x^2+15x+45$
2	Remove the '-5' (additive inverses sum to zero) so it now becomes	$7x^2+15x-5+5 = 5x^2+15x+45+5$
		$7x^2+15x = 5x^2+15x+50$
3	Remove the $15x$	$7x^2+15x-15x = 5x^2+15x-15x+50$
	or	$7x^2 = 5x^2+50$
4	Remove the $5x^2$	$7x^2-5x^2 = 5x^2-5x^2+50$
5	Divide both sides by 2	$2x^2/2 = 50/2$
	leaving	$x^2 = 25$
6	Find the square roots of both sides	$x = \sqrt{25}$
	RESULT	$x = \pm 5$

NOTE: The square root of any number should, strictly speaking, always be written as, for example, ±5. Only if we know from the circumstances that the value *cannot possibly* be negative are we safe in assuming that any square root is +ve.

11.16   Algebra and the Machine

11.16.1   The computer, unlike many of *us*, is not intimidated by a complicated model and could deal well enough with the original form of our equation but there are two problems.

There is the inefficiency (which ultimately slows program runs and costs money) of doing tasks which are unnecessarily complicated and, even more important, the greater risk of mistakes or increased error.

11.16.2   A second, but even more important concern, is people-related. *We* and any of our colleagues who may have to work on our programs, *must* be able to understand the models, not just when we have devised them but perhaps years afterwards.

It is then that the time taken in simplifying and standardising our models pays dividends.

11.17   Changing Forms

11.17.1   It happens very commonly that a model, even one which we have taken from a reference book, may not be in the appropriate form for the application which we have in mind.

In that event a certain re-modelling is needed.

11.17.2   If, for example, we consult a conventional maths book we shall find that the classic formula for the volume of a sphere is:

$$V = \frac{4}{3}\pi r^3$$

when V = volume and r = a *known* radius.

11.17.3   Supposing, however that we need to manufacture a sphere which will contain a specified volume and of which we must determine the radius, which is often the way in which the engineer meets the problem.

11.17.4   We then need to find an unknown 'r' for a *known* V, the exact opposite of the maths text book version. In algebraic language we must 'change the subject of the formula' or make 'r' the subject of the equation.

11.17.5   We do pretty much the same as the simplification process we have already considered, take one step at a time and use our knowledge of inverses.

1   Remove the 'divide by three' to give $3V = 4\pi r^3$
2   Divide both sides by 4 $3V/4 = \pi r^3$
3   Divide both side by π $3V/4\pi = r^3$

4   Find the 'cube root' of both sides $\sqrt[3]{\dfrac{3V}{4\pi}} = r$

11.17.6   Once again we see a standard algebraic method used to help us prepare our models in the form best suited, for a variety of reasons, to computer use.

11.18   Substitution

11.18.1   The final algebraic method with which we need to concern ourselves is that of SUBSTITUTION.

11.18.2   If, at any time whilst we are working on a model, we develop or discover actual *numerical values* for any of the terms we may often simplify matters still further by substituting the value(s) for the unknown(s).

11.18.3   In our most recent example, for instance, the use to be made of the solution will determine the appropriate value for π and when we know the former we should substitute the appropriate number in our model.

11.18.4   If the manufacturing demands are always the same we may be able thereafter to use a CONSTANT numeric value for π within our program.

11.18.5   Other substitutions may be a little less obvious at first glance. If we have a pair of simultaneous equations:

  i) $7x + 2y = 25$ and
  ii) $x + 3y = 9$

we could, by changing the subject of ii) conclude that $x = 9-3y$ and substitute $9-3y$ for any 'x' in equation i) to give: $7(9-3y) + 2y = 25$ and, by our step-by-step process find:

$$63 - 21y + 2y = 25$$
$$63 - 19y = 25$$
$$-19y = -38 \text{ **}$$
$$y = 38/19$$
$$y = 2$$

** When we have a minus quantity on each side it is often useful to multiply both sides by -1; can you see how this works?

11.18.6   The traditional manner of solving simultaneous equations might also be said to be a kind of substitution.

In the pair of equations we have just looked at we might substitute a version of equation ii) in which the 'x' component matches that of equation i), namely, $7x + 21y = 63$. We can do this by multiplying all the components of the original equation ii) by 7.

If we now set the equations together:

i) $7x + 2y = 25$ and
ii) $7x + 21y = 63$

we can see that the *difference* between the two equations, found by *subtracting* i) from ii) is that

$$19y = 38 \text{ hence } y = 2,$$

which is a conclusion we arrived at in the earlier substitution method.

In both cases if we substitute this value for 'y' in either equation we shall be able to find the other *unknown*, 'x' and check our results by making the substitution also in the other equation.

11.18.7   What we have done is to substitute for an *inconvenient* form of one of the equations a version in which, whilst better suiting our purpose, *the original equivalence relation is maintained*.

We may even make convenient substitutions of *both* equations if it suits our purposes.

11.18.8   Take for example the simultaneous pair:

i) $3p + 2q = 4$
ii) $4p + 5q = -11$

which will not reveal a usable *difference* by changing one equation only but will do so if we substitute other versions of *both*:

i) $12p + 8q = 16$
ii) $12p + 15q = -33$

leaving, as the difference,

$$7q = -49 \text{ or } q = -7$$

11.18.9   Check that you understand how the substitute equations were arrived at and try some examples.

11.19   The Computer and Simultaneous Equations

11.19.1   The method is, however, complicated to program for the computer since it involves making *judgements* about relationships between the equations, and computers are not good at inspecting equations and making judgements about the relationships between them.

11.19.2   Fortunately we have yet another long-standing device in the mathematics toolbox in which, rather than substituting another form of the equation(s) we set the values concerned into a matrix and then substitute the INVERSE of the MATRIX.

11.19.3   If we look at our first pair of equations we can write the whole thing in matrix form as:

$$\begin{bmatrix} 7 & 2 \\ 1 & 3 \end{bmatrix} \begin{pmatrix} x \\ y \end{pmatrix} = \begin{pmatrix} 25 \\ 9 \end{pmatrix}$$

11.19.4   What we have *not* seen hitherto is that not only do numbers and algebraic terms have inverses but so also do matrices, both additive and multiplicative.

For addition $\begin{bmatrix} 7 & 2 \\ 1 & 3 \end{bmatrix} + \begin{bmatrix} -7 & -2 \\ -1 & -3 \end{bmatrix} = \begin{bmatrix} 0 & 0 \\ 0 & 0 \end{bmatrix}$ shows the inverse law.

11.19.5   The multiplicative inverse of a matrix is a little more complicated, but is, happily, freely available in many translators and in virtually all spreadsheets systems. For any 2×2 matrix A we have the model:

$$\frac{1}{A} * A = \begin{bmatrix} 1 & 0 \\ 0 & 1 \end{bmatrix}$$

and we have seen, in an earlier chapter, both the *identity matrix*, with 1s in the leading diagonal and zeros in all the other cells, *and* the consequence of multiplying any array by it.

11.19.6   We now need a rule for finding the INVERSE of any 2×2 matrix to solve equations with two unknowns. If we label the cells in our 'subject' matrix a, b, c and d, we have:

$$\text{the inverse of } \begin{bmatrix} a & b \\ c & d \end{bmatrix} = \frac{1}{ad-bc} * \begin{bmatrix} d & -b \\ -c & a \end{bmatrix}$$

and it remains only to substitute the *numeric values* for any pair of equations with two unknowns into the inverse matrix in all the appropriate places.

11.19.7   The part *outside the matrix proper*, '$1/(ad-bc)$' is known as the DETERMINANT of the matrix and when we are finding an inverse by hand we must remember to evaluate the determinant, (*ad–bc*), and, as a final step, to use its *inverse* to multiply our product matrix.

11.19.8 In spreadsheets and translator programs all these calculations are invisible to us and are 'called' by the ARRAY INVERSE command in one form or another.

11.19.9 Let us try our example, multiplying both sides by the inverse matrix.

$$\frac{1}{21-2}\begin{bmatrix} 3 & -2 \\ -1 & 7 \end{bmatrix} * \begin{bmatrix} 7 & 2 \\ 1 & 3 \end{bmatrix} \begin{pmatrix} x \\ y \end{pmatrix} = \frac{1}{21-2}\begin{bmatrix} 3 & -2 \\ -1 & 7 \end{bmatrix} * \begin{pmatrix} 25 \\ 9 \end{pmatrix}$$

becomes

$$\frac{1}{19}\begin{bmatrix} 19 & 0 \\ 0 & 19 \end{bmatrix} * \begin{pmatrix} x \\ y \end{pmatrix} = \frac{1}{19} * \begin{pmatrix} 57 \\ 38 \end{pmatrix}$$

then

$$\frac{1}{19}\begin{bmatrix} 19 & 0 \\ 0 & 19 \end{bmatrix} * \begin{pmatrix} x \\ y \end{pmatrix} = \frac{1}{19} * \begin{pmatrix} 57 \\ 38 \end{pmatrix}$$

and finally

$$\begin{bmatrix} 1 & 0 \\ 0 & 1 \end{bmatrix} * \begin{pmatrix} x \\ y \end{pmatrix} = \begin{pmatrix} 3 \\ 2 \end{pmatrix} \text{ giving } x = 3; \ y = 2$$

11.19.10 This is the preferred method of dealing with such simultaneous equations by computer since, once the numbers are properly placed into an array, the normal matrix functions do all that is necessary.

11.20 Coefficients

11.20.1 It remains only to add one comment about *notation*. If you follow up any of these ideas by reading conventional mathematics textbooks the *number values* associated with symbols, for instance the '7' in $7x$ is known, in the jargon, as the COEFFICIENT of x.

11.20.2 Taking this one step further and looking at $7x^2$, for instance, the actual quantity is made up of *three* parts, the COEFFICIENT, the UNKNOWN (value of 'x' itself) and the INDEX (or POWER) of the value represented by the 'x'.

11.21 Conclusion

It is in ways like these, which all apply just as much to the MODELLING and solution of INEQUALITIES, that classical numerical and algebraic methods can help us with our practical modelling tasks and it is for this reason that at least the techniques and ideas we have looked at in this chapter are worth spending some time on.

Exercises

1) We can be sure that $1 < \sqrt{2} < 2$ since $1^2 = 1$ and $2 \div 2 = 1$. As a first estimate of $\sqrt{2}$ we could try 1.5.

   Squaring this gives 2.25, which is larger than 2 and so 1.5 must be too big an estimate.

   We can now say that $1 < \sqrt{2} < 1.5$.

   As an improved estimate we could now try 1.25 and squaring this gives 1.5625 which is less than 2 and so our new estimate is too small and $1.25 < \sqrt{2} < 1.5$.

   Work through this iteration until you can confidently write $\sqrt{2}$ correct to 3D.

2) Write a pseudo-code segment to perform this iteration and test it.

3) a) Devise an iterative model to solve $y = \sqrt{x}$ where 'x' may take any input value.

   b) How will your model differ if answers are needed to :

      i) 3D;
      ii) 4 sig. figs.?

4) Find $\sqrt{470}$ by an iterative method.

5) A vulgar fraction value for π is 22/7 and calculating this gives 3.143 to 3D. Use the iteration in 11.8.1 of Chapter 11 to calculate π to 3D.

   How large is the difference between the two values?

6) Write the model $P = R*H - T$ with, in turn, R, H and T as the subject.

7) a) Simplify $7x^3 + 2x^2 - 7 - y = 4x^3 - x^2 - 5$

   b) If $x \in$ {Integers: $x : 1 \leqslant x \leqslant 4$} calculate, for your simplified equation, the corresponding values of 'y'.

8) Write $y = 2x^2 + 27$ with x as the subject.

9) Solve the simultaneous equations:

   $3t + 2s = 27$
   $2t + 5s = 51$

a) by 'traditional' algebra;

b) by the inverse matrix method.

10) Express $\dfrac{(x^5 * x^{-1})}{(x^{-2} * x^4}$ in its simplest form.

12
Diminishing Uncertainty – Statistics

Objectives

After working through this chapter you should be able to:
- understand what we mean by the word 'statistics';
- see its relevance to the work of computer staff;
- perform elementary statistical analyses of sets of data;
- interpret elementary descriptive statistics.

12.1 Why Statistics?

12.1.1 There are many situations in the real world which cannot be modelled by any of the techniques we have considered so far. In the introduction to this book some such problem areas were identified as being those where:

a) certainty involves destructive testing;

b) the situation is too large or too complex to be measurable with any certainty;

c) we need, on the basis of what has happened in the past, to be able to make useful forecasts of what is likely to happen in the future.

12.1.2 These are typical of many of the information problems of the organisations which our computer systems serve where:

a) applies to products which it manufactures or uses;

b) relates to the organisation itself, the country within which it functions, or the markets in which it hopes to compete; and

c) concerns the monitoring of performance and profitability without which the management cannot *plan* but can only *react*.

12.1.3 All of these will also apply, to a greater or lesser extent, to our computer systems themselves.

12.1.4 The branch of mathematics which covers such modelling problems is generally known as STATISTICS.

12.1.5 This word means many different things to different people but for us it is essential to be very clear about the precise nature of what we are dealing with, because all our systems are likely to have to meet demands, from our users, for 'statistical information'.

We must understand those needs and know how to create or use the software which will produce the information demanded of us.

12.1.6 Within any substantial computer system we shall also need statistical tools to help us evaluate the capacity of those systems, to plan the work-load which they can carry and to appraise the reliability of existing equipment or the many choices available to us when we come to replace or extend any part of our systems.

Brochures for even the simplest type of printer may be full of real or pseudo 'statistical' information of which we *must* be able to appraise the validity, the truthfulness and its usefulness to us.

12.2 The Stages

12.2.1 Within what is generally described as 'doing statistics' there are a number of distinct phases:

1) planning the 'capture' of the raw data;
2) *capturing* the data;
3) organising that data into usable form;
4) displaying the data in ways which are clear, *true*, and easy to understand;
5) analysing the data by approved methods;
6) deriving *statistics* from the analysis;
7) stating the statistics in approved ways;
8) indicating what conclusions may be drawn;
9) identifying any further assumptions which may be INFERRED from the statistics.

12.2.2 You will notice that in stages from 1) to 5) I have spoken only of 'data' and that 'statistics' are actually not mentioned until stage 6). This is because statistics is like the information processing cycle where raw data, which is of little use until *processed*, is *input* to a system whose *output* is information.

12.2.3 In statistics we have exactly the same separation between the raw data and the statistics.

12.2.4 Strictly speaking, a STATISTIC is *the numerical output of a recognised process for evaluating raw data*.

People *do*, it is true, use the word 'statistics' very loosely in conversation and, worst of all, they say things like "Statistics prove . . . .".

12.2.5 This, like so many misinterpretations of the subject, is wholly untrue. Statistics may *describe* a situation or, at best, provide *evidence* as to the whether a given situation seems to be what is to be expected.

12.2.6 What can be *derived* from that evidence, stages 7) and 8), involve more complex tasks which are not our concern at this level. We are concerned with the tools of analysis and reporting, leaving the conclusions and inferences to the user.

12.3   Language

12.3.1 The language of this type of modelling deals with 'tendencies', 'trends' or even 'likelihoods' but *never* certainty; at best we can *diminish uncertainty*.

This does not imply that we may be casual or careless. We must learn to approach the subject in a strictly disciplined and cautious way but, happily, there is very little really difficult mathematics involved and, as always, when we meet any new symbols or models we will explore and explain them step-by-step.

12.3.2 We must also remind ourselves that both data and information may be QUALITATIVE or QUANTITATIVE. The former relating to *attributes*, and therefore to be dealt with by us, if at all, either by SETS or NON-PARAMETRIC statistical methods. In this book we are not concerned with the latter.

12.3.3 *Quantitative* data, which *are* very much our concern, are expressed in *numbers* as we saw earlier, but these, let us remind ourselves, may be on NOMINAL, ORDINAL or RATIO scales and the latter may be DISCRETE or CONTINUOUS. For the purposes of this chapter we shall limit our work to the *ratio* scales.

12.4   Sampling

12.4.1 Ordinarily the responsibility for stages 1) and 2), in our list above, will lie with the people who *design* the statistical investigation and the data which has been collected will be presented to us for processing.

12.4.2 What we *shall* need to know is whether the data are from a SAMPLE or from the whole POPULATION. For the NCC International Diploma, for example, the *population* is *all* the potential candidates in *all* the schools and colleges in *all* the regions of *all* the countries concerned.

12.4.3 The candidates of any one school will, on the other hand, be a *sample* of that population. For all sorts of reasons that sample may be TYPICAL or ATYPICAL (*not* typical).

That is to say that it may or may not be truly REPRESENTATIVE of the population as a whole.

12.4.4   For a truly reliable sample we may have to take our data at RANDOM from the whole population.

There are whole books on sampling methods if you wish to pursue this topic.

12.4.5   It is, however, very important to us when we are trying to appraise a manufacturer's claims for the performance of disc drives, for instance, to know how the statistical evidence was gathered.

12.4.6   Unrepresentative samples can generate statistics which give a very misleading impression because the data are, by intention or otherwise, *biased*, 'tilted' towards producing favourable statistics, and therefore must be considered to be defective.

Some of the models which we use will need to be modified if the data are from small samples.

12.5   Practical Steps

12.5.1   Let us start to model a practical problem as a means of exploring the stages from 3) to 6).

A stock-update run is to be re-scheduled and we need to decide where to fit it into the new plan. A sample of 72 runs over several weeks has been timed with the results, in minutes to 2D, below:

20.72	14.53	23.31	19.25	16.87	7.15	17.44	12.28	11.30
19.55	23.85	17.14	4.85	10.95	21.65	28.10	11.91	29.64
15.38	11.57	22.61	21.16	13.20	16.03	19.03	22.08	18.39
17.06	24.14	15.19	12.29	19.47	13.39	17.77	23.31	15.16
25.02	16.29	20.03	13.57	16.61	19.41	22.97	18.64	13.12
19.36	12.99	18.73	22.44	14.73	19.83	14.46	19.51	15.82
12.58	26.78	18.28	9.53	20.38	21.30	13.56	17.62	22.80
27.54	10.54	17.59	20.45	23.74	18.70	26.31	11.03	12.41

12.5.2   This is typical of raw data in that it is un-ordered and almost unreadable as it stands. Before we can interpret such material we need to know if it fits any 'pattern'.

12.5.3   We need to know both the minimum and the maximum times, or the RANGE, whether there is some sort of 'average' time and whether most runs are close to that time or are widely scattered.

12.5.4   In formal terms we speak of MEASURES of CENTRAL TENDENCY on the one hand and measures of DISPERSION on the other. (Note that we only use the word 'average' in a very general way, not to mean anything precise, since it has come to mean so many different things that it no longer has any useful meaning at all mathematically.) We shall explore these measures more fully as they become relevant.

12.5.5 With very small numbers of data, fewer than twenty or so items, we may simply TABULATE and ORDER the data.

12.5.6 With larger samples, in order to avoid the sheer volume of work in dealing with all the data separately and because the process often throws up other useful information, we group the data into CLASSES.

12.6 A Distribution

12.6.1 Here, in planning how to organise the data, we need to introduce some new techniques in order to *bring out* the pattern, or DISTRIBUTION of the data.

12.6.2 In order to decide on what classes to use we need first to establish the SCALE of measurement and the RANGE. In this case the scale is continuous and the range is from the smallest value, 4.85 minutes, to 29.64 mins. (This clearly *does* need further investigating since it would obviously be very foolish to schedule 4.85 minutes for every run of the job and very wasteful always to allow 29.64 minutes!)

12.6.3 It is of considerable practical value, in order to reveal the symmetry of the data, if it exists, to choose an ODD NUMBER of classes, ideally not fewer than five and not more than eleven, with INTEGERS as their mid-points in order to minimise error effects.

Our classes need to cover a range of at least from 4 to 30 inclusive, with a middle value of 17. *Seven* classes each having a 'spread', or CLASS INTERVAL, of 4 minutes, would cover a very slightly larger overall range and be convenient to work with.

12.6.4 It is also the case that our data imply a CONTINUOUS scale so that our classes must allow for this by defining boundaries precisely.

As a matter of notation we commonly use 'x' for an individual datum and 'x_i' to identify any *specific* value. We use 'n' for the total number of data.

Our suggested division would give:

$$3 \leq x < 7$$
$$7 \leq x < 11$$
$$11 \leq x < 15$$
$$15 \leq x < 19$$
$$19 \leq x < 23$$
$$23 \leq x < 27$$
$$27 \leq x < 31$$

Note carefully the use of the '<' and '≤' symbols and that the use of them unmistakably defines each class as a true set so that there can be no doubt as to the class for which any individual datum is intended.

12.6.5 We now do something which we would think quite improper in ordinary arithmetic models. We assume that all items which fall into a particular class are evenly scattered within that class and that the mid-value can represent any and all of them as a kind of 'average'.

12.6.6 Accordingly we do not record *any individual datum* but merely *count* the *number* of data which fall into each class. In practice, we put these classes into a table, make a mark for each item in a class and gather those marks into groups of five. We have, for centuries, called this 'tallying' items.

12.6.7 Can you write a short pseudo-code segment which could allocate the data to the appropriate classes and perform the 'tallying'?

12.6.8 The number of items in each class, after tallying, is known as the FREQUENCY of that class and the whole process of organising the data in this way is described as making a FREQUENCY DISTRIBUTION of the data. In Figure 12.1 I have, as the final step, added a column to contain the class frequencies.

CLASS	X^i	TALLY	f^i
$3 \le x < 7$	5	I	1
$7 \le x < 11$	9	IIII	4
$11 \le x < 15$	13	LHT LHT LHT II	17
$15 \le x < 19$	17	LHT LHT LHT LHT I	21
$19 \le x < 23$	21	LHT LHT LHT III	18
$23 \le x < 27$	25	LHT III	0
$27 \le x \le 31$	29	III	3
		Total = n =	72

Figure 12.1 Tallying the data

12.6.9 To sum up this phase we have made a *frequency distribution* of our raw data by allocating them to seven *classes* with a *class interval* of four, the *midpoint* of each class being taken as the *value* of each item in that class and with the total 'tallies' being the *frequency* for each class.

12.7 Displaying the Data

12.7.1 We can now put the relevant data into a spreadsheet, for example, and DISPLAY these data in pie charts, bar, or line graphs if all we wish is to see the picture more clearly. The *display* of such material is, for the great majority of people, the best way of making it easily understandable.

12.7.2 Here, in Figure 12.2, is the simple *histogram* of the data and this already tells us a certain amount which is useful.

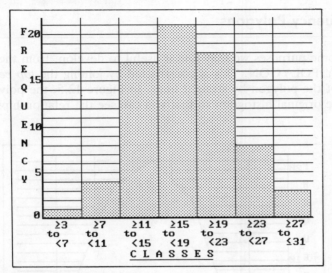

Figure 12.2 The straightforward histogram

12.7.3 For example it is clear that *most* of the runs occupy the three centre columns and that a very few took very much less, or very much more time than the majority.It gives a good immediate picture of the data.

12.7.4 We should pause at this point and look at the overall picture presented by the histogram, which may be broadly symmetrical, as in our example, or may be 'skewed' either 'negatively' or 'positively, as shown in Figure 12.3 and Figure 12.4, below.

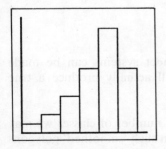

Figure 12.3 Negative skew

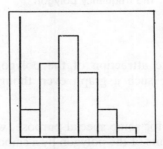

Figure 12.4 Heavy positive skew

12.7.5   Symmetry is a norm for many distributions and if we see a histogram with a heavy skew, in either direction, we should call attention to the fact as part of our report.

12.7.6   There may well be good reason for the lack of symmetry which the user will either understand or will wish to investigate further.

For our present purposes we should not attempt to *interpret* such results.

12.8   Frequency Polygons

12.8.1   For some purposes we may find it useful to develop, from the histogram, a FREQUENCY POLYGON. This is constructed by joining the *top centre points* of the columns of the polygon by straight lines. In Figure 12.5, the polygon has been overlaid on the original histogram so that you can see the relationships.

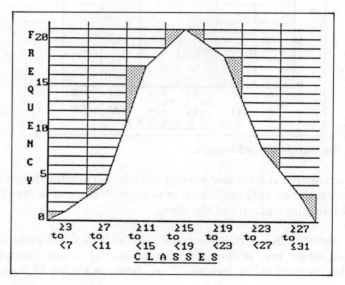

Figure 12.5   The frequency polygon

12.8.2   One attraction of the polygon is that spreadsheet systems can be made to generate such a graph even though few, if any, will actually produce a true histogram.

12.8.3   If, for some special purpose, we use a very large number of classes we may join the points of the polygon by a smooth curve. It is often such 'smoothed' polygons which are shown as models of certain 'standard' distributions.

12.9 Cumulative Frequency

12.9.1 The information given, either by the histogram or the polygon is, however, primarily IMPRESSIONISTIC rather than QUANTITATIVE and to this extent the graphs, whilst very useful, are still rather limited.

There is a good deal to be gained by adding another new technique and tabulating the CUMULATIVE FREQUENCY, often written as 'Cu.f', of the distribution, either in ABSOLUTE or in RELATIVE terms.

Table 12.1

Mins.	f_i	ABSOLUTE Cu.f.
<7	1	1
<11	4	5
<15	17	22
<19	21	43
<23	18	61
<27	8	69
<31	3	72

12.9.2 From Table 12.1 we can see that 1 run took *less than* 7 minutes; a total of 5 runs took less than 11 minutes; 22 runs in all took less than 15 minutes, and so on.

This, useful though it may be, would not tell us much if we were comparing with a performance plan or with *another* sample of say 97 runs and, as is usual when we wish to compare two or more ratios or proportions, it is the common denominator of PERCENTAGES to which we turn.

12.9.3 The *absolute cumulative frequencies* are converted to percentages to produce the *relative cumulative frequencies*. In our present example 1/72 runs took less than 7 minutes; 5/72 took less than 11 minutes; 22/72 took less than 15 minutes, and so on. Converting these to percentages we can both simplify and extend our table as shown in Table 12.2:

Table 12.2

Mins.	f_i	ABSOLUTE Cu. f.	RELATIVE Cu. f.
<7	1	1	1.4
<11	4	5	6.9
<15	17	22	30.5
<19	21	43	59.7
<23	18	61	84.7
<27	8	69	95.8
<31	3	72	100

12.9.4   The contents of the table may now be plotted on a graph as in Figure 12.6.

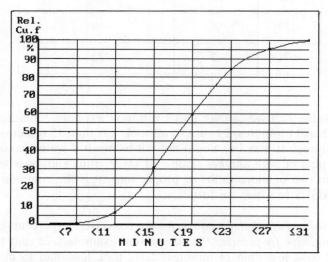

Figure 12.6   A cumulative frequency ogive

12.9.5   This curve does not resemble any of those at which we looked in the chapter on graphs but is known as an OGIVE. The complete graph, plotted from our *relative cumulative frequency* data is known as the CUMULATIVE FREQUENCY OGIVE of the distribution. In practice, we rarely use absolute frequencies for this purpose so you may often find that the 'relative' part of the title has been left out.

12.9.6 If the vertical scale is to 100% it is *relative*, if numeric, as the *actual frequencies* and the maximum is the 'n' of the data, it is *absolute*.

12.9.7 As it stands this graph does not tell us much more than the histogram but we *can* read from it our first *measure of central tendency*.

12.9.8 In any distribution the actual MIDDLE VALUE is often of interest and we have come to call this the MEDIAN.

12.9.9 On the frequency ogive the 50% point identifies the median and if we 'drop' a perpendicular line from where the curve itself is cut by the 50%, horizontal line, we can read off the *value* of the median on the scale of the VARIABLE, '*x*', which is 'minutes', in our present case.

12.10 Quartiles

12.10.1 We also often divide a distribution into equal 'slices' of the cumulative frequency. Among the commonest of such sub-divisions is to identify the 25%, 50%, 75% and 100% levels of cumulative frequency, dividing our distribution into quarters.

We have already seen that the 50% mark leads us to the *median*; similarly the values at the 25% and 75% points on the frequency scale will lead us to corresponding points on the horizontal scale.

12.10.2 Values of the *measured variable* which are derived from such regular sub-divisions of the frequency scale are called by the rather strange name 'fractiles'. If we divided the frequency scale into sixths, for examples, the corresponding values of the variable parameter would be 'sextiles'.

When we subdivide the '*f*' scale into quarters the resulting values are called QUARTILES.

12.10.3 From this 'annotated' version of the ogive we can now read both the *median*, our first *measure of central tendency* and the FIRST and THIRD QUARTILES, representing the 25% and 75% frequency 'bands' respectively. This can be seen in Figure 12.7.

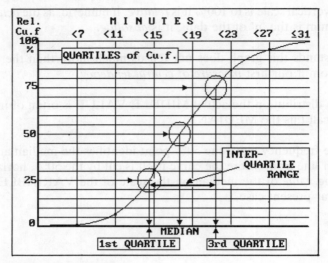

Figure 12.7   The annotated ogive

12.10.4   We can now say that 25% of runs took not more than 14 minutes and that 25% took more than 21 minutes, with both values being approximate since they are read from the graph.

12.10.5   To reach the highest level of approximation in the interpretation of ogives we must, once again, use large scales on accurately printed graph paper and learn to draw the curves as accurately as we possibly can. Nowhere is pride in good craftsmanship better repaid than in the drawing of graphs.

12.11   Disperson or Spread

12.11.1   We can also see that 50% of cases took between 14 and 21 minutes, This is the range between the first and third quartiles and gives us our first useful measure of spread, or DISPERSION, which we call by the rather clumsy name of the SEMI-INTER-QUARTILE RANGE, the SIQR, or, more briefly, the QUARTILE DEVIATION.

12.11.2   We find this by *halving the absolute difference between the first and third quartiles*. In our current example the SIQR $\approx \frac{1}{2}|14{-}21|$ or 3.5 = the quartile deviation.

12.11.3   In statistics the word 'deviation' has two special meanings. It is either the amount by which any datum differs from the 'average', or it is one of several kinds of *'average' of the spread of individual values about a central point.*

12.11.4 What is important about this technique of classifying the data and drawing the ogive is that, once the range has been allocated to classes, there is no further special formula or elaborate model to build. We can simply read off the two measures from the graph.

12.11.5 Moreover, the whole technique is ideally suited to working by spreadsheet up to the point of plotting the *points* on the graph. It is better to print out the graph *then* and draw the curve by hand since spreadsheets systems, as we saw earlier, will ordinarily not draw smooth curves.

12.11.6 We can read many other answers from the ogive. Suppose, for instance, the Operations Manager decided that the schedule, ideally, should allow 20 minutes for this particular job run. How often would it 'over-run'? Look at 20 on the 'minutes' axis, draw a line up to the curve and then across to the frequency scale and we see that approximately 70% of runs would be completed within time and 30% would run over time. He or she might well think that this was an acceptable percentage of overruns. The 'worst case' scenario will be when *all* the jobs happen to over-run on the same shift!

12.11.7 The next benefit from preparing such a graph and deriving the two measures is that by doing so for each of two samples, even of quite different sizes, we can compare medians and SIQRs to see if they are closely similar, or markedly different performances. This is the great importance of STANDARD MEASURES of central tendency and dispersion, that we can *compare* sets of data with each other or with STANDARD MODELS.

12.11.8 The cumulative frequency ogive is a quite powerful model which is easily derived, yet from which we can obtain much useful information. It must be seen, nevertheless, as a relatively limited device and there are other and more sophisticated measures which suit more demanding tasks.

12.12 Central Tendency

12.12.1 If we look back once again at our *histogram* we see that there is one column which represents a greater frequency than any other, the $15 \leqslant x < 19$ class. This class, which contains the largest number of items, is often important and within it lies the MODE, a second measure of central tendency.

12.12.2 For a reasonable approximation, we may take the mid-value of the class as the mode but, if we wish, we may use a slightly more refined technique. In this we take some account of the relative heights of the two adjacent columns by drawing two lines, as shown in Figure 12.8, and dropping a perpendicular to the horizontal scale.

The value indicated by the arrow, between 15 and 19 on the scale, is the mode but it is clear that the graph, as with the ogive, must be drawn as precisely as possible if we are to be able to read such values to a sufficient degree of accuracy to justify the extra complications.

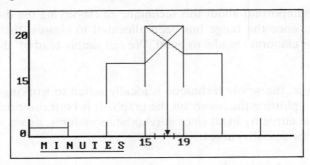

Figure 12.8   The mode refined

12.12.3   Before leaving this topic we must be aware that distributions may be BI-MODAL, may have two modes and appear something like Figure 12.9.

Once again this is an output to which we should call attention, but should not attempt to interpret. The user may well expect such an outcome or may, if, is unexpected, mount an appropriate investigation. We are not, at the present level, competent to draw the correct inferences.

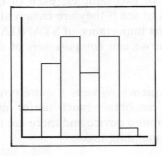

Figure 12.9   A bi-modal distribution

12.13   Full Descriptions

12.13.1   One of the limitations of our *quartile deviation*, as a STATISTIC, is that it describes only the middle 50% of a distribution and ignores the outer 50% which is spread over the two 'tails', as they are called. Since there may be considerable variation between the tails of this and any other distribution, there will be many purposes for which we need more detailed information which takes account of the *whole* of the data.

12.13.2   We need to look at a more detailed description (and what we are dealing with are all known as DESCRIPTIVE STATISTICS, often available under statistics software packages by the command 'DESCRIBE'). It is most convenient to do this by taking a simpler data set, to keep the arithmetic to a minimum, because we shall be using some new techniques and building a new model which can then be developed for the more elaborate data sets.

12.13.3 We need, let us remember, a measure of *central tendency* and a measure of *dispersion* for a data set like : 12, 9, 7, 18, 5, 14, 11, 19, 15, 8. This is clearly too small to put into classes and so we deal with it in a different way. First we find the 'average', in this case the ARITHMETIC MEAN, by adding all the *x*s and dividing by the number of items. This is the *third* of our common *measures of central tendency* which are the MEDIAN, MODE and ARITHMETIC MEAN. Each is arrived at by a different procedure, describes different aspects of the distribution and is used for different purposes. This is why we treat the word 'average' with such caution. There are at least *five* commonly used measures of central tendency and what the layman means by 'average' might be any one of them!

12.13.4 This is an opportune moment to meet a very common but rather forbidding-looking use of symbols. We may well find, in a mathematics or statistics textbook, the arithmetic mean indicated by :

$$\bar{x} = \sum_{i=1}^{i=n} x_i \div n$$

Let us take this, symbol by symbol and translate it into plain words.

The symbol $\bar{x}$ designates the ARITHMETIC MEAN of all the values of x whilst the 'n' is the number of data items, which we have met before. The unfamiliar Greek upper-case 'S', the Σ symbol, means 'add together all the values' of whatever is to its right, in this case 'x_i', the individual values of our data variable.

Underneath the Σ is '$i=1$' which means '*from x_1*' whilst *above* the Σ is '$i=n$' which means 'to the 'nth' value of x', or in other words the symbols are mathematical shorthand for, "The arithmetic mean (of a set of data) is found by adding together all the values of x from the first to the last and finally dividing by the number of data items".

This is all that the elaborate picture really means but it is valuable because it saves quite a lot of words, is completely clear to any mathematician or statistician anywhere in the world, and is wholly un-ambiguous. Like so much mathematical symbolism it is a very powerful shorthand and the use of such symbols is especially common in statistics.

In practice the 'from' and 'to' parts of the notation are often left out and it is written as 'Σx_i' but the meaning is exactly the same. If, however, we wish to sum (add together) only *part* of a data set we *must* use the full version.

12.13.5 For brevity, once again, we commonly use only the word 'mean' when we are speaking of the *arithmetic mean* but we must be extremely careful here since that is but one of several 'means' which are in use.

Perhaps we should agree that we will use 'mean' *only* when we intend the *arithmetic* version and that we will use the full name of any other 'means' which we may encounter.

12.13.6 Once we have established the mean, we can calculate the DEVIATION of each datum, that is to say *the amount by which it differs from the mean*, which we usually label '*d*'. We must recognise that some values will deviate to the *lower* side of the mean and some to its *higher* side. Those below will, then, have negative values whilst those above will be positive. It is a good idea always to show the *sign* of the individual deviation since it is often a matter of considerable importance. Being below 'average' is *very* different from being above.

12.13.7 Let us prepare our data table so far:

Table 12.3

n=10	x_i	d_i
	12	+0.2
	9	−2.8
	7	−4.8
	18	+6.2
	5	−6.8
	14	+2.2
	11	−0.8
	19	+7.2
	15	+3.2
	8	−3.8
Σ	118	0.0
Mean x̄	11.8	

12.13.8 To find the 'average' of the deviations is clearly impossible since $\Sigma d_i = 0$. This is inherent in the very nature of the arithmetic mean and will always be the case.

In order to make all the deviations into +ve quantities, so that we may find a mean, we resort to a very old mathematical trick and *square* each of the deviations. If we then find Σd_i^2 we *can* find *its* mean and finally take the *square root* of that mean in order to arrive at a true mean of *all* the deviations, not just half of them as in the *mean deviation*.

12.14 Standard Deviation

12.14.1 Some of you, if you have done much work in electrical or electronic theory, may recognise this process as finding the *root mean square*, or RMS of a set of data. For our purpose in statistics we call it the STANDARD DEVIATION and to describe it we use either the abbreviation S.D. or the Greek letter 'σ'. Usually we shall see the formula:

$$\sigma = \sqrt{\frac{\Sigma (x_i - \bar{x})^2}{n}}$$

Our amended data table will be as seen in Table 12.4:

Table 12.4

n=10	x_i	d_i	d_i^2
	12	+0.2	0.04
	9	−2.8	7.84
	7	−4.8	23.04
	18	+6.2	38.44
	5	−6.8	46.24
	14	+2.2	4.84
	11	−0.8	0.64
	19	+7.2	51.84
	15	+3.2	10.24
	8	−3.8	14.44
Σ	118	0.0	197.6
Mean x̄	11.8		σ≈4.45

12.14.2 In this case we find that the *mean square* of the deviations is 19.76 and this value is often used in statistics and is called the VARIANCE or σ^2.

12.14.3 Here we meet a new problem and a new remedy. It has long been established that small samples, below about 40 items (some authors suggest 50) tend to produce results with substantial inherent error. In order to diminish this we apply a correction, known as a Bessel function after the mathematician who devised it, by using, for finding σ or σ^2, '$n-1$' rather than 'n' in all our calculations so that the formula becomes:

$$\sigma = \sqrt{\frac{\Sigma (x_i - \bar{x})^2}{n-1}}$$

which would give, for our present set of data, *a variance* of 21.96 and a *standard deviation* of 4.69.

12.14.4 It is not usual to do statistics with very small *populations* and it is wise to use the '$n-1$' version of all statistical formulae when dealing with situations where $n < 40$. In many texts such small data sets will be called 'ungrouped' data, such as we have just looked at.

12.14.5 Data which has been set into classes is often referred to as 'grouped' data.
   Let us now look again at our set of grouped data and see how that affects the calculations of $\bar{x}$ and α.

For our variation on the S.D. model we change the detail of the tabulation to avoid having first to find $\bar{x}$ and the individual deviations.
   As you will recall, we take the mid-value of the class as the 'x' for all members of that class. Our new tabulation provides the means of using that simplified procedure.

Table 12.5

x_i	f_i	x_i^2	$f_i * x_i$	$f_i * x_i^2$
5	1	25	5	25
9	4	81	36	324
13	17	169	221	2873
17	21	289	357	6069
21	18	441	378	7938
25	8	625	200	5000
29	3	841	87	2523
	n=72	Σ	1284	24752

12.14.6   First let us see a different arrangement of the model which we used earlier to find σ. It looks more complicated but is not really greatly so.

$$\sigma = \sqrt{\frac{\Sigma\,(f_i x_i^2)}{n} - \left(\frac{\Sigma f_i x_i}{n}\right)^2}$$

Within this model we find, inside the right-hand pair of brackets, the mean, which we earlier did manually. This value, when squared, is subtracted from the mean of the *total* 'x^2' values, on the left.

Substituting in this formula the values in our table, above, we find that this gives $\bar{x} \approx 17.83$ and $\sigma \approx 5.07$ for the full data set.

12.14.7   Compare these with the MEDIAN and SIQR which we found by the cumulative frequency model earlier. Can you account for any differences there may be?

For ease of manipulation this is by far the simplest model since, from the time of deciding the classes and their mid-points, and the subsequent tallying of the data to produce the frequencies, all is simple arithmetic (if rather tedious and subject to human mistakes).

The model may, with a little care, be used in a spreadsheet system without amendment and the spreadsheet system can perform all the tedious arithmetic.   If we need to use it at all frequently, it lends itself ideally to being manipulated by a *macro* written in the meta-language of the spreadsheet. It is not even too difficult to write a program segment to perform the task, although there are many suitable packages on the market.

12.14.8   The mean, median, mode and standard deviation are the major DESCRIP-TIVE or SUMMARY STATISTICS offered by any good statistics software package. The choice as to whether to use the tabulation/cumulative frequency ogive technique, or the *calculated values* which we have just been looking at will depend upon what use is to be made of the statistics and whether speed is more important than detail.

12.15   Error

12.15.1   Before leaving these descriptive statistics there is a last important concern. Virtually all numbers have inherent error, as we saw in Chapter 5. All our raw data, then, will have such error and the *statistics derived from them will also have error*. Not only are $\bar{x}$ and σ subject to induced error, rounding error and the like, but they will be affected by the accumulated error within the data and induced by the processing.

12.15.2   We need not go into the calculations which lead to the concept of STAND-ARD ERROR but we do need to know how to calculate and publish such results from our processing. The two main ones, for our purposes are:

i) the STANDARD ERROR OF THE MEAN S.E. $_{\bar{x}} = \dfrac{\sigma}{\sqrt{n}}$

ii) the STANDARD ERROR OF THE S.D. S.E. $_{\sigma} = \dfrac{\sigma}{\sqrt{2n}}$

12.15.3   For many purposes we should append these to the reported $\bar{x}$ and σ respectively when we make our final printout.

12.16   Trends

12.16.1   One very valuable tool to our users is a is a TREND ANALYSIS and we may well be asked to produce the analysis of data which will allow the necessary graph to be generated. Typically, the data will be describing something like this:

Table 12.6
Quarterly Stock-holding Figures in $000

YEAR	1st QTR.	2nd QTR.	3rd QTR	4th QTR
1988	32.0	32.7	32.5	33.9
1989	35.3	34.2	34.0	35.8
1990	37.5	35.9	36.8	38.2
1991	39.4	37.9	38.3	37.9

There are, it is clear, many fluctuations and we may see these on any ordinary graph. What we need, however, is an indicator of whether the variations are RANDOM, and therefore UNPREDICTABLE or whether there is some underlying pattern.

12.16.2   One tool which we use in modelling this situation is known as the MOVING AVERAGE, (in this case the 'average' is the *arithmetic mean*).

12.16.3   If the data are in strict sequence of TIME as in this case, the collection (32.0, 32.7, 32.5, 33.9, 35.3, 34.2,......37.9) is known as a TIME SERIES. (In strict mathematical terms it should be known as a 'sequence' *not* a 'series' since it does not obey the rules governing the latter but the name is the common one and we must accept it.)

What follows, however, is that both *time* and *stock level*, in this case, are variables, each seemingly on a discrete ratio scale but with an *implied continuous scale*.

Table 12.7

YEAR	QUARTER	s_i	$\bar{x}_i$
1988	1	32.0	
	2	32.7	32.35
	3	32.5	32.42
	4	33.9	33.16
1989	1	35.3	34.23
	2	34.2	34.22
	3	34.0	34.11
	4	35.8	34.95
1990	1	37.5	36.23
	2	35.9	36.06
	3	36.8	36.43
	4	38.2	37.32
1991	1	39.4	38.36
	2	37.9	38.13
	3	38.3	38.21
	4	37.9	38.06

This is to say that, although the stock was actually valued at specific points on the time scale and appears to have discrete values, the *actual* stock value will have been varying on a day-by-day basis and between any two quarters will probably have fluctuated considerably.

12.16.4 Since we now have *two* values which are changing, as compared with one variable and *fixed* class intervals in earlier models, we have a BI-VARIATE distribution as distinct from the UNI-VARIATE data which we have seen hitherto in this chapter.

12.16.5   Table 12.7 shows what develops if we 'map' the series contained in our data table in 12.16.1 onto the 'moving average' by the model:

if s_i = any individual stock level
and $\bar{x}_i$ is any individual average then:

$$\bar{x}_1 = \tfrac{1}{2}(s_1 + s_2); \quad \bar{x}_2 = \tfrac{1}{2}(\bar{x}_1 + s_3); \quad \bar{x}_3 = \tfrac{1}{2}(\bar{x}_2 + s_4) \text{ and so on.}$$

Note that, in order to minimise error propagation, the $\bar{x}$ values are to 2D and that there is no $\bar{x}$ for the first quarter of the time series.

12.16.6   We may now plot the data on a graph (shown in Figure 12.10) and since the 'average' may be assumed to be moving *continuously*, not in *discrete* steps, as are the stock figures, the latter should be shown as a *broken*, the former as a *continuous*, smooth line. If we wish we may now attempt to draw a LINE OF BEST FIT, that is to say a line which, by eye, seems to 'average' the continuously varying *moving average* line on the graph. I have drawn in such a line in the graph of our data and we see three distinct patterns. First we see the quite large fluctuations in stock value on the quarter-to-quarter basis, shown by the broken lines.

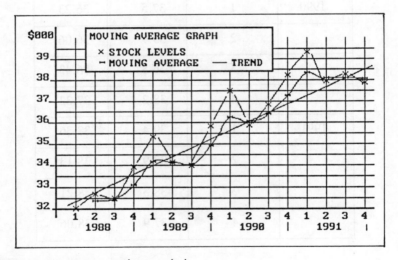

Figure 12.10   A time-series graph extended

12.16.7   We next see the 'smooth' variation of the moving average as a continuous but still fluctuating function, but varying much less than the 'point' stock values. Finally, we have the straight line on the graph, which is an attempt to interpret the moving average fluctuations as an overall TREND.

12.16.8   Although estimated, this gives a useful indication. We must not, however, set too much importance on what we might *extrapolate* from this trend since we have no evidence upon which to make any confident predictions.

12.17 Scatter

12.17.1 The final type of analysis of data which concerns us here is where not only do we have two variables but it is reasonable to assume that one *might* vary in some sort of relationship to the other.

12.17.2 In such circumstances, we call the one which we believe might be the *cause* the INDEPENDENT VARIABLE which we always plot on the *horizontal axis* of a graph with the other, the DEPENDANT variable, plotted on the *vertical axis*.

We may assume, for example, that people's weight depends to some extent upon their height. It is clear that if this is so it is not the greater *weight* which causes people to grow taller. It might, on the other hand, be reasonable to suppose that as we grow taller with the passing years our weights also increase. In this case the *independent variable would be height* with weight being the dependant. We do need to be a little cautious when deciding between any two variables and it is worth pausing for thought.

12.17.3 Suppose we assume that the sale price of shops dealing in the same goods will largely depend upon their weekly takings and that a sample gives the data in Table 12.8:

Table 12.8

SHOP	TAKINGS $00 ($x_i$)	SALE PRICE $000 ($y_i$)
A	10.0	25.6
B	6.0	12.4
C	5.5	10.2
D	8.0	14.0
E	16.5	38.5
F	7.5	13.0
G	12.7	27.0
H	15.0	34.8
I	9.5	14.0

We can tell little from looking at the table other than that there seems to be a tendency for the price to rise as the takings get larger, but to be rather more sure we need to see the picture more clearly. Whenever you find yourself thinking like this, it is well worth while to try plotting a graph.

12.17.4 In this particular case the individual values of the variables, whilst on continuous ratio scales, are actually not connected to each other since each shop is quite a separate entity from all the others. We can only, in this situation, plot a POINT GRAPH.

For each shop we plot the point at which the takings and the price correspond and we may, as in our example, label each of the points.

12.17.5 What we now look for is any *pattern* to the scatter of the individual points.

12.17.6 In Figure 12.11, the points all lie close to the 'line of best fit' which has been put in 'by eye'. This suggests that the relationship between takings and the sale price of these shops *are* very closely linked, that they 'relate' to each other and we call such a correspondence the CORRELATION between the data.

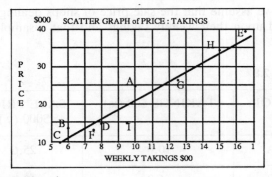

Figure 12.11 The scatter graph

12.17.7 Such scattered points may be quite clearly random, as in Figure 12.12, or may congregate within a circle and in such cases we may see that there is no correlation. The variables are not linked by some 'cause and effect' relationship.

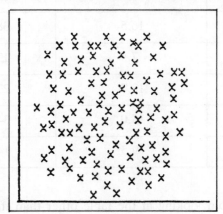

Figure 12.12 No pattern

12.17.8 If, on the other hand, the points fall within a close ellipse which slopes *upwards from left to right* (as shown in Figure 12.13) it indicates a strong *positive* correlation indicating that as the '*x*' value grows, so does the '*y*' parameter and that the correspondence between them is very close.

The 'fatter' the ellipse the weaker is that relationship.

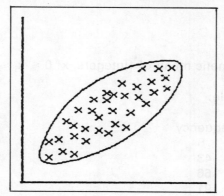

Figure 12.13 A strong +ve correlation

12.17.9 It is not likely to come as a surprise that a 'tight' ellipse which slopes from top left to bottom right indicates a strong *negative* correlation.

This means simply that the '*y*' parameter *decreases* as the '*x*' parameter *increases*. This is shown in Figure 12.14.

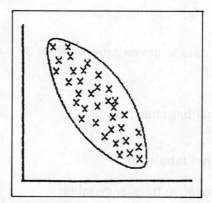

Figure 12.14 A strong −ve correlation

12.18 Conclusion

Once again we find powerful, ready-made techniques and procedures from the mathematical toolbox but we should be a little careful when we use them. We are, at this level, concerned with the *modelling* of the different statistical situations and should be very wary of inferring too much from the results.

That is a later stage of statistical analysis.

Exercises

1) What is the arithmetic mean of {Integers, x: $0 \leqslant x_i \leqslant 50$} ?

2) Here is a data table:

Age	Frequency
26–30	5
31–35	43
36–40	66
41–45	91
46–50	87
51–55	105
56–60	83
61–65	80
66–70	25
71–80	14
>81	1

a) Re-allocate the data to seven groups, ≤30; 30 – 40; 40 – 50; and so on. Show the data as:

   i) a pie chart;
   ii) a proportional bar chart;
   iii) a histogram.

b) Using the original tabulation:

   i) create a cumulative frequency table;
   ii) draw a relative cumulative frequency ogive;
   iii) state the median, mode and SIQR.

c) Calculate the mean, variance and standard deviation.

3) The turnover of a small shop in five successive months is:

   $18400, $17800, $21200, $17000, $15800.

   Can we assume, on the basis of the last three month's figures, that the business is declining?

256

4) An examination showed the following distribution of marks:

Marks	Frequency
0 < 10	3
10 < 19	5
19 < 28	5
28 < 37	8
37 < 46	12
46 < 55	17
55 < 64	24
64 < 73	17
73 < 82	6
82 < 91	2
91 –100	1

a) Draw the histogram.

b) Find the mean, mode and standard deviation.

c) There is a rough rule that in such circumstances 2/3 of candidates should have marks which fall within 1 s.d. of the mean. Does that hold good here?

d) A similar rule suggests that about 4% of candidates should have marks which differ from the mean by more than twice the s.d. How do these data compare?

5) A commodity traded in both Singapore and London showed the following average prices over an eight-month period:

SINGAPORE $ 83   85   88   80   95   98   97   94
LONDON £   28   29   29   27   31   32   32   32

a) Plot the data on a suitable graph.

b) Is there any sign that the prices in the two places are related to each other?

6) 59½, 60½, 61, 61½, 62, 62, 62½, 62½, 63, 63, 63, 63½, 63½, 63½, 64, 64, 64, 64, 64½, 64½, 64½, 64½, 64½, 65, 65, 65, 65, 65½, 65½, 65½, 65½, 65½, 65½, 66, 66, 66, 66, 66, 66, 66½, 66½, 66½, 66½, 66½, 66½, 66½, 67, 67, 67, 67, 67, 67, 67, 67½, 67½, 67½, 67½, 67½, 67½, 68, 68, 68, 68, 68, 68, 68½, 68½, 68½, 68½, 68½, 68½, 69, 69, 69, 69, 69, 69½, 69½, 69½, 69½, 69½, 70, 70, 70, 70, 70½, 70½, 70½, 71, 71, 71, 71½, 71½, 71½, 72, 72, 72½, 73, 73½, 74½.

a) Organise these data into an appropriate distribution and derive all the common descriptive statistics.

b) Produce any graphs which you think would help present a clear picture of the data.

13
Will it Work? Probabilities

Objectives

After working through this chapter you should be able to:
- understand the fundamental ideas of probability;
- use probability models to evaluate computer systems;
- model probabilities in a variety of ways;
- select models to suit specific needs.

13.1 Probability Modelling

13.1.1 The world is full of *un*certainties: there are in fact very few total certainties. Despite this we have to plan; we have to evaluate, and allow for, risks; we have to compare the reliability of systems and we have to attempt to forecast what is likely to happen.

Mathematicians have, over several centuries, developed and studied PROBABIL-ITY MODELS. It is to these that we turn in order to represent an uncertain world and to satisfy the practical demands of living in it.

In this respect the subject matter of this topic is closely allied to that of the previous chapter.

13.1.2 We, as practising computer staff, need to have a grasp of both the principles and the practice of the application of probability models and theories for two reasons.

Firstly our users, especially at management level, will need to make use of probability models for much of their planning and forecasting.

13.1.3 Secondly, at both the practical and the theoretical levels, our own computer systems are 'complex states', as referred to in the introduction to this book. They can only be evaluated by statistical methods and their behaviour forecast by probability models.

13.2   Language

13.2.1   There is, as always, some new language to learn but little in the way of new symbols.

The first word to have a special use is EVENT, to refer to anything which has happened in the past or may happen in the future. It is the all-purpose word for everything from tossing a coin to the printer going down in the middle of a run. The former is so trivial that we might hardly consider it worth calling an 'event' in real life; the latter may have such an effect on the workplan as to be seen as a catastrophe! Nevertheless, for our present purpose, each is simply an 'event'.

13.2.2   The second commonly used word is OUTCOME, to mean the result of an experiment or of a sequence of events. Roughly speaking it means something like 'final result'.

13.2.3   A third word is TRIAL, which indicates any individual test or example within a series of tests.

13.3   The Probability 'Scale'

13.3.1   A principle of major concern, from the outset of this study, is that the PROBABILITY of any EVENT occurring in the future is evaluated on a scale from 0 to 1.

13.3.2   The totally impossible event, (although it is very difficult to make a list of examples since scientific progress has a habit of making yesterday's certainties obsolete), has the probability 0.

13.3.3   Any absolutely certain event, again very difficult to list, has a probability of 1. Most events will have a probability, 'p', of $0 < p < 1$.

13.3.4   By whatever numerical process we evaluate past events and from them calculate what is likely to happen in the future our 'probability' answers cannot possibly be less than 0 or more than 1. Many, many marks are lost in examinations when candidates forget this basic rule.

13.4   A Simple Model

13.4.1   At the simplest we may look at a *single* event and evaluate it. Suppose that our printer, according to the records of the past year, has a mean downtime of 20 minutes per 8 hour shift, what is the probability that we shall get through a full shift with a working printer?

13.4.2 First we must compare like with like, which means that the same set or the same *measuring scale must be used*. Minutes are not, in the strict sense, like hours so we must state the recorded mean as 20 minutes out of any 480 minutes.

The probability that the printer *will* break down is 20/480 and the probability that it won't is 460/480. We often use 'p' to denote the probability that the event in which we are interested *will* take place and 'q' to indicate the probability that it *will not*.

13.4.3 The first 'law' of probability is that p + q = 1 and *no individual situation can be otherwise.*

In the present example, if we first reduce our fractions to 'lowest terms' we see that:

a breakdown has p = 1/24 = 0.0416. = 4.2%

whilst a shift unhindered by printer breakdown has: q = 23/24 = 0.9583 = 95.8%

and p + q = 24/24, 1.0 and 100% respectively.

The different ways of writing the fraction are all valid and equally commonly used so we should use the form which best suits the data.

13.4.4 NOTE here that we must not confuse probability with 'odds'. The odds *against* the printer breaking down, ie that it *will not*, are 23 to 1.
Be sure that you can see why.

13.4.5 These statements of probability, whilst being mathematically accurate, do not, of course, in any way *influence* the *actual* behaviour of the printer, although this appears to be the belief of many devoted gamblers when it comes to betting! Nor do they indicate what *will* happen on any *particular* shift; they refer only to the *risk* of breakdown.

13.4.6 We deal not in predicting certainty but in forecasting what is *likely* to happen. The Operations Manager or Shift Leader must either *know* the true risks, and take account of them in planning the workload, or just trust to luck that the system will be adequate.

13.5 Probability Spaces

13.5.1 In a more complex situation it is often useful to use a VENN diagram to model a PROBABILITY SPACE. We can then identify *all* the elements and include them in the universe of discourse. To take an example:

In a computer department of 120 staff there were 15 systems analysts, 40 programmers, and 56 operators. Of these, however, 24 of the operators were also competent programmers and 2 were also trained analysts; 10 of the programmers were also trained analysts and 1 person was competent in all three skills.

13.5.2   From this information we can construct a model which embodies all possible dispositions of skills.

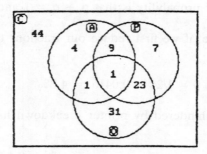

Figure 13.1   One form of probability space

13.5.3   The Venn diagram allows us to list, by direct observation, *all* relevant initial probabilities.

In the examples below we can list the probabilities as to what skills any member of staff, chosen at random, possesses.

13.5.4   Let us call our unknown person 'X'. The probability that he or she will be a programmer, an element of set 'P' is:

40/120 = 1/3 = 0.3 recurring = $33\frac{1}{3}$%

By using formal set symbols we can complete a list:

p (X ∈ A) = 15/120 = 1/8 = 0.125 = $12\frac{1}{2}$%
p (X ∈ O) = 56/120 = 7/15 = 0.46. = $46\frac{2}{3}$%
p (X ∈ (A ∩ O)) = 2/120 = 1/60 = 0.016. $16\frac{2}{3}$%
p (X ∈ (P ∪ A)) = 45/120 = 3/8 = 0.375 = $37\frac{1}{2}$%
p (X ∉ (X (A ∩ O))) = 118/120 = 59/60 = 0.983 = $98\frac{1}{3}$% and so on.

13.5.5   Can you complete the table for all *eight* domains of the diagram?

13.5.6   It is clear that we could also identify the probabilities for all the various combinations of *complements* and subsets and you should try a number of these.

13.5.7   There is *no* probability concerning the known data which is not retrievable from the diagram. If we, for instance, reverse the question and say "What is the probability that 'X' will *not* be a programmer?" we have a ready-made procedure. Since *all* the probabilities must total 1 the answer to our new question is 1 – p(X ∈ P) and this is 1 – 40/120 = 80/120 = 2/3 = 0.6 recurring = $66\frac{2}{3}$%.

13.5.8 If we take the histogram of the distribution of job-runs from the previous chapter, even this allows us to assess probabilities.

If, for example, we need to know how likely it is that any job taken at random will be completed in 13 minutes or less we can see that of the 17 jobs in that class half, or $8\frac{1}{2}$, are likely to be below the 13 minute point to add to the 5 in the two lower classes to give a total of $13\frac{1}{2}$ out of a total of 72 job runs in all which will take ≤13 minutes.

This is easiest to deal with as:

$27/144$ or 0.1875 or $18\frac{3}{4}\%$,

which is the probability which we are seeking. (See Figure 13.2.)

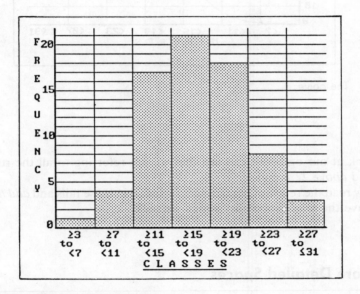

Figure 13.2 The histogram as probability space

13.5.9 You may already have concluded that the smooth curve of the cumulative frequency ogive, if well plotted and drawn, will actually give us many more answers than the histogram.

Check this statement by carefully plotting the curve for yourself then taking various *times* and evaluating the associated probabilities. Now look at Figure 13.3.

13.5.10 In the last chapter the 'tails' of a distribution were mentioned and by looking at both tails of the graph we can answer such questions as:

"What is the probability that any given run will be *more or less* than five minutes 'outside' the median?"

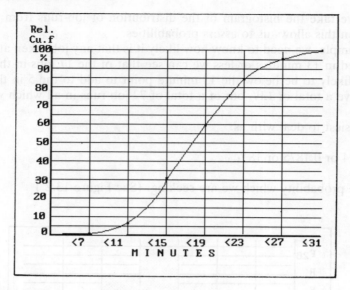

Figure 13.3 The ogive

If we look at our ogive we can see that we are referring to all the runs *below* 12 minutes and *above* 22 minutes.

If we now read the relevant percentages included in these tails and *add them together* we shall have the *total* probability which we seek.

13.6 More Detailed Spaces

13.6.1 The first of these questions, about this distribution, involved a ONE-TAIL, the second a TWO-TAIL test.

13.6.2 The usefulness of such tests of probability within the probability space represented by a frequency distribution highlights, once again, the difficulties of comparing *absolute* statistics.

13.6.3 If, for example, we wish to compare the performances of a person whose marks in Computer Studies are 67% and in Programming are 52% we have an apparently clear superiority in the first subject. This is quite deceptive since what we do *not* know is the relevant central tendencies and spreads of *all* the candidates.

If it turned out that the *statistics* were:

Computer Studies	$\bar{x} = 55; \sigma = 14$
Programming	$\bar{x} = 42; \sigma = 8$

we can see that the *overall* patterns of the two sets of results were very different and it is difficult to see whether either of our candidate's performances is better or worse than the other.

13.6.4 *Theoretically*, if a 1 – 100 scale is being fully used, we should expect the mean to be 50% and the S.D. 14 marks.

However, different numbers of people sit different tests with different examination papers. Different markers are subject to all sorts of human vagaries as are the people who sat both tests in our example. So many differences in circumstances lead to differences in the statistics, such as those in our example, being commonplace. They *do*, however, prevent any meaningful comparisons on an *absolute* basis.

13.7 Standard Scores

13.7.1 We must therefore find ways of making *relative* comparisons and, in order to be able to do so on a useful basis, we have the concept of the STANDARD SCORE (or MEASUREMENT) denoted by the symbol '*z*'. We find '*z*' for any individual data item, x_i, by the formula:

$$z_i = (x_i - \bar{x}) / \sigma.$$

13.7.2 Calculate the '*z*' scores for our candidate's two performances. Your results should be:

Computer Studies	$z = +0.86$
Programming	$z = +1.25$

In both cases his or her performance was above the mean but the performance in Programming was actually better, compared with the overall performance of all candidates, than was the performance in Computer Studies.

13.7.3 When we are seeking probabilities of any given event within a *large* population, we commonly use a probability distribution model which is a smoothed frequency polygon scaled in STANDARD (*z*) SCORES.

13.7.4 In the illustration in Figure 13.4 is shown the curve of the 'normal' distribution, which is characteristic of a great many situations, such as human heights – or performance in Computer Programming examinations, if the total population is *very* large.

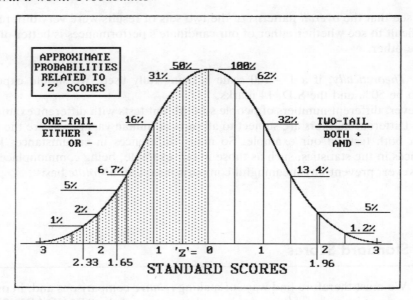

Figure 13.4 The 'normal' curve

13.7.5 The graph has been subdivided into one-tail and two-tail versions of the same 'z' scores scale and the statistics around the perimeter of the curve show the proportion of cases *outside and including* that particular 'z' score.

We can see that, for example, approximately 16% of cases will have 'z' scores of more than 1 (standard deviation *above* the mean). If we look for those who have scored more than 1 s.d. *less than the mean* we find that this is also approximately 16% of the total.Each of these, *individually*, is a one-tail test.

13.7.6 If we ask for *all the cases* which differ from the mean by at least 1 s.d., whose 'z' scores are ≥1, as a complete group both above *and* below the mean, we shall find the answer on the *two-tail* side of the graph and it is, not surprisingly, 32%.

13.7.7 From this pattern we can establish the *probability* of any individual 'z' score arising.

If, for example, we look back at our candidate's Programming test performance, which had a 'z' score of +1.25, we can see, on the one-tailed side of the curve, that such a 'z' score describes somewhere between 16% and 6.7% of candidates.

13.7.8 To find out *just what that percentage is* we may use the technique of LINEAR INTERPOLATION.

1.25 is half-way between 1 and 1.5 so the percentage of cases concerned will be *half-way* between 6.7 and 16, namely 11.35. This means that only 11.35% of candidates, in a normal distribution, would reach or exceed a 'z' score of +1.25.

The performance in Computer Principles, on the other hand with a 'z' score of +0.86 represents some proportion of the population between 0.5 and 1, or between 31% and 16% of the population. the difference in 'z' scores is 36/50 of the gap between

0.5 and 1 so the value we are seeking will be 36/50 of the gap between 16% and 31% or: $31 - 36/50(15) \approx 20.2\%$. The performance reached by *that* 'z' score is that which might be expected of that proportion of the population.

13.7.9 Rather than going to the trouble of *calculating* such interpolations, if we are likely to be much concerned with statistical probabilities, we shall find a table of 'z' values and the corresponding proportions of the normal distribution in any statistics textbook or good mathematical tables.

13.7.10 Putting these results into the form of *probabilities* is now very straight-forward, since the probability in each case is the same as the proportion of the population which might be expected to reach that score.

13.7.11 On the graph have been marked the 5% and the 1% PROBABILITIES of any 'one-tailed' event.

13.7.12 By the process of reading the graph 'in reverse', so to speak, we find that there is only a 5% probability of any such event having a 'z' value of $\geqslant 1.65$ and only a 1% probability of it having a 'z' value of $\geqslant 2.33$.

We can also use this information the opposite way round, so to speak.

13.7.13 On the two-tailed side, there is only a 5% probability of events falling into the combined tails above or below $z=1.96$. This means that for a large number of events which might be expected to conform to this 'normal' distribution, the probability is that 95% will fall within the band of $\pm \leqslant 1.96$ s.ds from the mean.

13.7.14 The normal curve is, like any other frequency distribution, another kind of probability space within which are to be found all the evaluations which we might wish to use.

13.8 Multiple Probabilities

13.8.1 So far we have looked only at *single* events but many of the situations with which we have to deal involve a *number of elements* which may influence outcomes.

A typical computer system, for example, will incorporate a processor and a number of peripherals. Each of these elements will have its own speed of data transfer; its capacity for handling or storing data; as well as its rate of failure, both as we have logged them in action *and* as the manufacturer's *claimed* parameters for all of these attributes.

In planning the procurement of systems, in evaluating their *actual performance* and what workloads may consequently be planned for, we are, inevitably, dealing with probabilities and statistics.

13.8.2   Before considering actual cases, we had better spend a little time looking at simplified situations so that we may establish some rules.

13.8.3   Every individual event is, in one sense, completely new. We sometimes sum this up in the saying, "A coin has no memory." This means that no matter how many times we toss a coin every *new* toss is a completely fresh start.

13.8.4   In any sequence of trials, however, events may be INDEPENDENT or COMBINED. In the former case, any one event has no influence on or relation to any subsequent event. In other cases we may be concerned with quite complex *trains* of events.

13.8.5   Every single independent event has a determinable probability for both 'p' and 'q'. In the sake of a coin p = ½ and q = ½.

13.8.6   When we consider events which may be *combined*, we take their independent probabilities and combine them arithmetically in one of two ways. If the events are combined in an AND relationship the independent probabilities are MULTIPLIED to find the outcome. If the association of events is by an OR relationship, the independent probabilities are ADDED. Initially, these may be represented by BOTH and EITHER respectively.

13.8.7   We may see, from an earlier example, the BOTH and the EITHER *relationships* although, since the members are already allocated to their sets, no further arithmetic is needed.

In Figure 13.5 we see, shaded, the 10 people who figure in the A *and* P set.

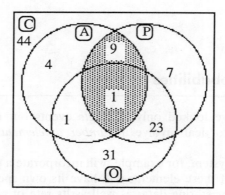

Figure 13.5   In *both 'A' and 'p'*

In Figure 13.6 it is the people who are *either* in A *or* P. We should note particularly that because of the intersections the *number* of people concerned is not 15/120 + 40/120 but 45/120.

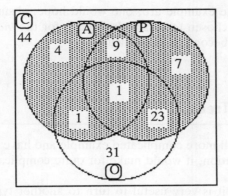

Figure 13.6   In *either 'A' or 'p'*

13.8.8   To look first of all at just the arithmetic involved, let us consider the case of a computer system which used two disk drives whose recorded histories gave their respective probabilities of failure as:

$$p_1 = 1/10 \quad \text{and} \quad p_2 = 1/15,$$

the probability that *both* would fail at the same time would be:

$$p_1 * p_2 = 1/10 * 1/15 = 1/150.$$

13.8.9   The converse, that both *would not* fail at the same time, is

$$9/10*14/15 = 126/150.$$

The sum of these two is clearly *not* 1 so what has gone wrong?

13.8.10   The answer is that there are other possibilities, that one would fail and the other not. This can happen as either

$$9/10*1/15 \ or \ 1/10*14/15.$$

13.8.11   Can you see why? If we now add *all the possibilities* we find

$$\frac{1+126+9+14}{150}$$

which, happily, equals 1.

13.8.12   The probability that EITHER one or the other would fail at any given time would be:

$$p_1 + p_2 = 1/10 + 1/15 = 3/30 + 2/30 = 5/30 = 1/6.$$

(The data used in the example are, one would hope, absurdly large failure rates for disk drives but are chosen to be simple enough for the *method* to be made clear without the arithmetic becoming too difficult.)

13.9   Probability Trees

13.9.1   If we take a still more complicated example and have *three elements* or more in our system configuration, it would make for quite complicated arithmetic.

13.9.2   It is then that it is very useful to turn to another type of model which we have met before.

13.9.3   Suppose, still using much simplified numbers, that we have a CPU, a VDU, a disk drive and a printer, with probabilities of failure of 1/20; 1/15; 1/50 and 1/10 respectively.

There will be many possible combinations of these four elements.

13.9.4   To define fully the probability space we may use a tree model. In the example shown in Figure 13.7 the layers are the pieces of equipment, in turn, and the paths are uniformly pointing upwards if the item will work and downwards if it will not.

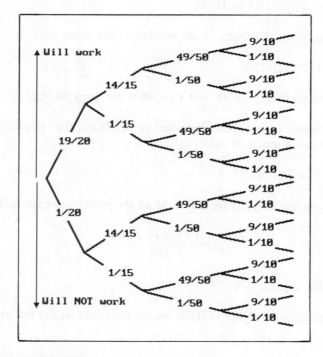

Figure 13.7   The vulgar fraction tree

13.9.5 On each path is the probability of each of the possible *independent* events and the complete tree embodies the whole of the probability space. Any combination of probabilities may be read off from the tree.

13.9.6 For example we have, as the probability that *all four* items will work, the sequence of four consecutive upwards-pointing paths.

Since this is an AND sequence we have, as the final outcome,

$$p = 19/20 * 14/15 * 49/50 * 9/10,$$

all dissimilar fractions and to multiply them we have to do considerable number 'crunching'.

13.9.7 We must have an eye for accumulated error in such circumstances but it may be that this is a case where the common base of *decimal fractions* will serve us better and we can state them as:

$$p = 0.95*0.93.*0.98*90.$$

The answer, 0.782, indicates that there is slightly better than a 78% probability of a trouble-free system at any given time, even with such poor reliability as our specimen probabilities indicate.

13.9.8 It is often the case that properly *calculated* probabilities are quite different from what we may have expected, so it is wise always to double-check our results.

13.9.9 Now take all possible paths through the tree and put the probability of each 'and' outcome as the 'leaf' at the end of each path. What *should* the total of all the leaves be? What total do you arrive at?

13.9.10 From the same tree the probability that, for example, EITHER the processor OR the VDU OR the printer is not working at any given time will be $1/20 + 1/15 + 1/10 = 13/60$ or $21\frac{2}{3}\%$.

13.10 Other Experiments

13.10.1 There remains only one other complication for us to explore within this topic, and we can see this most clearly in the context of some traditional practical experiments.

13.10.2 If we place three white counters and three black counters into a bag and, whilst blindfolded, remove one counter, what is the probability that it will be white?

It is not too difficult perhaps to see that

$$p = 3/6 = 1/2.$$

If we *keep out* the counter we have taken, which *did* happen to be white, what is the probability that a second 'draw' will also produce a white counter?

We have only 5 counters left, of which 2 are white so p_2 is 2/5. The probability of first a white counter AND a white counter second is 1/2 * 2/5 or 1/5, 20%.

13.10.3   Now repeat the experiment slightly differently by *putting the first counter back* before drawing the second. Both independent probabilities would now be 3/6 or 1/2 and 1/2 * 1/2 = 1/4 = 25%. There *is* a difference and quite an important one between the two conditions.

13.10.4   In our original test the probability of taking *either* a white *or* a black counter is 1/2 + 1/2. Since this equals 1 the outcome is certainty. This seems self-evident since the counter *must be one or the other*; there *is no other possible outcome*.

13.10.5   If we put 3 white, 3 red and 4 black counters into our bag the probability of taking out *either* a white *or* a red counter will be

$$3/10 + 3/10 = 3/5.$$

13.10.6   We can go on to 'compound' such experiments almost indefinitely with questions like, "What is the probability of finding either a red or a white counter on the first trial and then either a red or a black counter on the second trial?"

The arithmetic becomes a little more involved and we may need to take it one step at a time until we are more familiar with such models.

13.10.7   The first 'either' outcome is 3/5 and the second is 7/10 (by the same additive process if the first counter were replaced in the bag, but 7/9 if it were not).

The two 'eithers' are, however, joined in the question by an 'and' which means that the two 'partial outcomes' must now be multiplied together: 3/5 * 7/10 = 21/50 or 42%.

13.10.8   This is another type of situation where, if we need to deal with more complex problems, any good probability text will show us the necessary procedures.

13.11   Conclusion

There is, fortunately, much in quite elementary organisation of data which allows us to derive probabilities of events without extremely elaborate procedures and some of our existing tools serve us very well in this respect.

Some theoretical knowledge is, however, helpful in deciding which tools to use in which kinds of situation.

Exercises

1) From within a computer staff of 20 people, four are to be selected for a project team of whom one is to be the team leader.

   a) In how many ways could the choice be made?

   b) If the team were selected at random what is the chance that one particular person will be in the team?

   c) If a brother and sister work in the department what is the likelihood that:

   i)   both will be selected;
   ii) one of them will be in charge?

   d) If the first selection were made and then cancelled so that a further selection had to be made on the next day, each choice at random, what is the probability that brother and sister will each be chosen once only, but not on the same day?

2) Five of a computer staff of 22 are programmers.

   a) If you walk into the department what is the probability that the first person you meet will be a programmer?

   b) Does your answer to a) change and if so how, if you are a programmer in that department.

3) From the data in example 2 in the exercises for Chapter 12:

   a) what is the probability that any individual person will be:

   i)   not older than 40;
   ii) more than 80 years old;
   iii) between 50 and 60 years old?

   b) if any three were selected at random what is the probability that none of them would be over 60 years old?

273

4) In a set of examinations 60 candidates gained marks which showed the following statistics:

	MEAN	S.D.
Maths	48	16
Programming	55	12

a) If one candidate scored 80 for Maths and 65 for Programming what were his or her 'standard scores'?

b) The percentage of candidates normally scoring more than given standard ('z') scores is shown in this table:

'z' score	0.4	0.6	0.8	1.0	1.2	1.4	1.6	1.8	2.0	2.2	2.4
% scoring higher	34.5	27.4	21.2	15.9	11.5	8.1	5.5	3.6	2.3	1.4	0.8

i) What is the probability of each of the scores in a) ?
ii) How many people out of those sitting might be expected to have done better in each subject?
iii) What is the probability of anyone gaining this combination of marks?

14

Applied Two-state Logic and Switching

Objectives

After working through this chapter you should be able to:
- understand the importance of two-state logic in computing;
- model logical relations by truth-tables, Venn diagrams, switching circuits or 'gates';
- analyse and simplify circuits by Boolean algebra;
- use Karnaugh maps.

14.1 Logic in Programming

14.1.1 Sooner or later, in all but the simplest of programs, we meet a situation where we have to write something like, "IF *so-and-so* THEN DO *something*, ELSE DO *something different*."

This means that we *must* be able to model the problem in two-state logical terms, where the circumstances are so closely defined that the "IF" can only be answered by *YES or NO*; there may be no doubt.

Such absolute statements are quite foreign to ordinary discourse between humans which is full of approximations and shades of meaning.

14.1.2 It is for these reasons that failures of *logic* within our programs are just as common as failures of the syntax of the language we happen to be using and often far more difficult to detect and eliminate. Indeed, if our logic *is* in any way defective it may not become apparent until long after the program has gone into use when some unusual set of circumstances or piece of data, which we had not foreseen or tested for just happens to come along.

14.1.3 The *logic* of a program is entirely *independent* of the programming language used and we should be able to construct our models in pseudo-code and test our solutions properly, before final coding, to make sure that they do precisely what is intended.

The situation is often complicated by the fact that we may be making several logical decisions within a single line of code. If, for example, we are sorting data into the classes which we used for our statistical distribution of the run-times, early in Chapter 11, we might well do so by a line such as:

IF x<7 THEN $c_1 \leftarrow c_1+1$ ELSE IF x<11 THEN $c_2 \leftarrow c_2+1$ ELSE IF x<15 THEN $c_3 \leftarrow c_3+1$ ELSE IF x<19 THEN $c_4 \leftarrow c_4+1$ ELSE IF x<23 THEN $c_5 \leftarrow c_5+1$ ELSE IF x<27 THEN $c_6 \leftarrow c_6+1$ ELSE $c_7 \leftarrow c_7+1$.

14.1.4   This involves COMPOUND decisions within which *each single decision* must be *logically* correct, without any doubt whatever.

14.1.5   This is the realm of RIGOROUS or TWO-STATE LOGIC, which we have already met, but in a slightly different kind of application. We refer to it also as BINARY LOGIC and in the textbooks we use '1' for 'TRUE' or 'YES' and '0' for 'FALSE' or 'NO'. For technical reasons, which you may be able to relate to the work done in Chapter 6, many processors use '−1' instead of the '0' for NEGATION, as the 'no' is often called.

For simplicity we will follow the '1 / 0' convention in this book.

14.2   Logic and Hardware

14.2.1   Apart from this fundamental importance in programming we shall also need to know precisely how the *logical* two-state system maps on to a *physical* two-state switching system if we become involved with assembly languages or the theory and architecture of the computer, or if we need to program any process control applications.

14.2.2   Today such systems, from controlling traffic lights to running the cycles of a washing machine, embody the logical models in various arrangements of miniaturised transistor switches in 'chips'. In this context, 'yes' ordinarily translates as 'switch *on*' and 'no' as 'switch *off*'.

14.3   Language and Symbols

14.3.1   We saw in our chapter on SETS how the strict mathematical conditions which govern them allow us to define *all* the relations in a given universe of discourse in a Venn diagram. We shall make further use of such diagrams in this chapter but, in spite of their very great usefulness and the symbolism of unions, intersections and the like, they do have limitations.

Try your hand at drawing a Venn diagram with four or five intersecting sets where each pair of sets has an intersection and where there is a subset embracing *all* of them. You will, I think, see how much more difficult it becomes.

14.3.2 The Venn technique is also much more powerful for evaluating *enumerated* sets than it is for solving more complex problems about the very basic relationships between sets.

We shall, therefore, need to investigate the two-state logical world a little more deeply and in such further study there is, as always, some different language and some new symbols to be learnt.

14.3.3 Among the jargon words we meet 'STATEMENT' in a new usage. It is applied *very strictly* in two-state logic to mean any *single* proposition which can *only* be identified as either TRUE or FALSE.

"He is male" is a statement which satisfies these new criteria, "He is a man" is *not*. This is because the set of males *is* defined but the age at which a boy becomes a man is not. Hence the second sentence involves what is *arguable* and is not, therefore, a statement.

14.3.4 Unlike the material of the previous chapter, the realm into which we now venture admits *only* of total certainty. That certainty associates 'yes' with 'true' and binary 1; 'no' with 'false' and binary 0.

14.3.5 Statements may be COMPOUND, when we associate two or more by the *conjunction* AND, or by the *disjunction* OR. "He is male and he is a programmer", is a CONJOINT compound of two statements. It is only *ultimately* true if *both parts* are true.

"He is a policeman or a thief" is a DISJOINT compound statement which is ultimately true if *either OR both* part(s) is/are true.

It is worth looking at these more carefully so as to get the foundations absolutely secure. The first tells us that a male may be *not* a programmer and that a programmer may be *not* a male, since, as we saw in the chapter on sets, every IS has a corresponding IS NOT. The *compound* statement is true, however, only if the 'he' in question IS both.

The second compound tells us that 'He' may be *not* a policeman or *not* a thief, but the compound statement is true if 'he' is *either*.

14.3.6 We often use the small letters p, q and r as the symbols for this sort of logical statement and we clearly need symbols for 'not', as well as 'and' and 'or'. In our first example we might allocate 'p' to 'is male' and 'q' to 'is a programmer' and we would write the compound statement as 'p ∧ q' to mean 'the consequences' of the two statements in *conjunction*.

I say *consequences* because there are *four* possible results of any compound of two statements, as we shall see a little later.

We would write, of our second compound statement 'p ∨ q' to indicate the possible results of the *disjunction* of p,q. The 'not p', or 'it is false that he is a policeman', is symbolised by ~p.

14.4   Truth Tables

14.4.1   By now the ideas and the symbols may have begun to cause some confusion so let us use our primary tool for simplifying such models, the TRUTH TABLE.

14.4.2   There are, for any two statements in compound, four possible relations. Any 'p' may be *only* either true or false and any 'q', likewise, may have only these two states. This leads to a table such as that shown in Table 14.1:

Table 14.1

p	q	p ∧ q
True	True	True
True	False	False
False	True	False
False	False	False

The *disjunction*, on the other hand, gives Table 14.2:

Table 14.2

p	q	p ∨ q
True	True	True
True	False	True
False	True	True
False	False	False

14.4.3   Each statement has its own TRUTH VALUE and the outcome of the compound statements is dictated by the ways in which the individual truth values are associated. The tables contain *all possible results* of such associations.

We can see the compound relations, at this very elementary level, in Venn diagrams which, if we look a little more carefully, reveal the need for a *third* compound. This is shown in Figure 14.1.

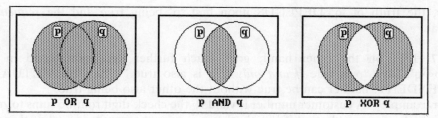

Figure 14.1 The 'or', 'and' and 'xor' domains

14.4.4 When we use the word 'or' in everyday life it is often ambiguous. "These seats are reserved for elderly or disabled persons" reads a notice in a 'bus: "Admission half-price for Children or Senior Citizens" reads a poster on the side of the same 'bus.

Can you see the important difference between the uses of the word 'or'? If you are in any doubt look again at the Venn diagram.

The first statement permits of people being *either* elderly or disabled or *both*, whilst the two groups in the second statement are *mutually exclusive*; no-one can be *both*.

14.4.5 In logic we distinguish between OR, meaning in either of the named sets *or* their intersection, and XOR, (exclusive or), which means in either *but not both*. There *is* a symbol for XOR, $\veebar$, but it is not universally accepted and is less reliable than the others.

The truth table for XOR is:

Table 14.3

p	q	p XOR q
1	1	0
1	0	1
0	1	1
0	0	0

14.4.6 All such statements, in programming, are set in an 'IF . . . THEN' program line but the word 'if' is also ambiguous in ordinary speech and needs clarifying for our purposes.

In some programming languages we find both IF and the 'IFF' which has been invented in order to eliminate the ambiguity.

The former describes a situation, symbolised by 'p→q', which says "if p is true then q is also true".

Or, the truth of 'q' DEPENDS upon first 'p' being true and the *sequence* is critical.

14.4.7 IFF, on the other hand, goes much further and, symbolised in turn by 'p↔q', says "p is true *if and only if* q is also true." They are MUTUALLY DEPENDENT; neither can be true without the other also being true.

For example, "A customer number is valid IF the check-digit routine sums to zero," certainly does *not* imply that if the checksum is zero the number is valid; there will always be cases, although low in probability, where the checksum will still be zero if there has been a peculiar transposition of digits. Yet we quite commonly find even experienced programmers and analysts making such assumptions, but it is merely an 'IF', not an 'IFF'.

"The customer may have a 5% discount IF his account remains within his credit limit," is fundamentally different. It is an 'IFF'.

If he is within his credit limit he may have the 5% discount but if he receives a 5% discount he *must* be within his credit limit. Programming and testing such differences is not always easy.

14.5   Logical Equivalence

14.5.1 The use of truth tables leads also to another condition, the LOGICAL EQUIVALENCE of compound statements. If the outcomes of two truth tables are identical, the statements which they represent are *logically equivalent* and we use the '≡' symbol to represent this state.

14.5.2 For example, let us look at $\sim(p\wedge q)$ and $(\sim p \vee \sim q)$

Since the 'outcome' columns of each section of the table are exactly identical we can safely say that:

$$\sim(p\wedge q) \equiv (\sim p \vee \sim q).$$

We commonly present such tables as a direct comparison as shown in Table 14.4:

Table 14.4

p	q	p∧q	~(p ∧ q)	p	q	~p	~q	~p ∨ ~q
1	1	1	0	1	1	0	0	0
1	0	0	1	1	0	0	1	1
0	1	0	1	0	1	1	0	1
0	0	0	1	0	0	1	1	1

14.5.3 Such programming constructs as we have been discussing are known as CONDITIONALS and it is important, in many programming applications, that we should test them rigorously by verifying their truth tables.

14.5.4 The compound statements will often include more than two propositions and the table may then become quite large.

For example "If p and q and r then . . . " will generate Table 14.5 in which, for each of the *four* possible results of combining 'p' and 'q' there will be *two* possible conditions of 'r' giving:

Table 14.5

p	q	r	p$\wedge$q$\wedge$r
1	1	1	1
1	1	0	0
1	0	1	0
1	0	0	0
0	1	1	0
0	1	0	0
0	0	1	0
0	0	0	0

14.5.5 If the compound statement were "p and q *or* r" we might find that the interpreter we are using *automatically* takes the 'and' before the 'or', but is often safer to use brackets whilst we are working on the model and write it as (p$\wedge$q)$\vee$r and to insert an extra column into our table to accommodate the intermediate result:

Table 14.6

p	q	(p∧q)	r	(p∧q)∨r
1	1	1	1	1
1	1	1	0	1
1	0	0	1	1
1	0	0	0	0
0	1	0	1	1
0	1	0	0	0
0	0	0	1	1
0	0	0	0	0

Once again we see all possibilities and our testing of such a model, in program form, must include all the combinations which can give rise to the eight possible results.

14.6 Switching Diagrams

14.6.1 It has long been the practice to illustrate such possibilities, in CONTROL SYSTEMS, by reference to diagrams showing the switching ON and OFF of an electric light, as shown in Figure 14.2. If you were not very good at science at school don't worry, just try following the track of the diagram with a pencil. If you can get from one 'supply' point to the other without meeting any gaps the condition is 'ON' and the lamp will light. Otherwise it will not.

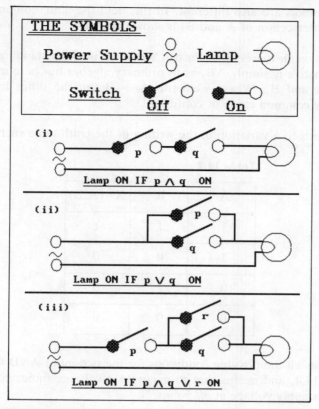

THE SYMBOLS

Power Supply   Lamp

Switch

Off   On

(i)

Lamp ON IF p ∧ q ON

(ii)

Lamp ON IF p ∨ q ON

(iii)

Lamp ON IF p ∧ q ∨ r ON

Figure 14.2 Three switching circuits

14.6.2 The switching function is carried out, within the computer, by circuits known as LOGIC GATES, for which statements have to be converted to electrical voltages, still identified as '1' or '0' for 'on' and 'off' but with individual *statements*, or INPUTS, labelled as A, B, C, and so on whilst *consequences*, or OUTPUTS, are labelled X, Y, Z.

14.6.3 These 'gates' are represented in diagrams as in Figure 14.3 but, just as with a truth table, compound statements may give any number of 'inputs' and gates may be combined in many different ways.

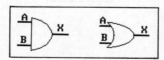

Figure 14.3 The two principal logic gates

14.6.4   The symbols are also different. In this field the 'and' conjunction of statements (or the intersection of A and B) is shown as A.B.

14.6.5   The dot *on the line between the two 'variables'* is a long-standing multiplication sign. The alternative is simply AB, as in ordinary algebra but here meaning, not A *times* B but 'A *and* B'. Where we write A + B, on the other hand, we mean 'A *or* B', a very common cause of confusion!

14.6.6   This leads to a variation in the writing of the truth table such as:

Table 14.7

A	B	X=(A.B)
1	1	1
1	0	0
0	1	0
0	0	0

This represents all the possible conditions for the two-input AND GATE symbolised in Figure 14.4, and means that to generate a voltage, indicated by '1', at the *output*, we must apply voltage at *all inputs*.

Figure 14.4   The and gate

14.6.7   The *disjunction*, on the other hand, uses the 'OR' gate (see Figure 14.5) and is represented by its own different version of the truth table, still with A, B and X, but using the '+' symbol for the 'OR' relationship.

Figure 14.5   An 'or' gate

In this case a voltage, (1), at the output is generated by an appropriate voltage at *any one or more inputs*. The table follows, below:

Table 14.8

A	B	X=(A+B)
1	1	1
1	0	1
0	1	1
0	0	0

Each table, once again, contains all the possibilities about the gate concerned.

14.6.8 The XOR gate is shown in Figure 14.6 for comparison.

Figure 14.6 The xor gate

14.6.9 If we wish to look at *combinations* of gates, as they represent more elaborate sequences of statements, we can start with the equivalent of the third of the light switching circuits in Figure 14.2, earlier.

In Figure 14.7 we see that our original 'p' and 'q' are represented by 'A' and 'B' and are inputs to an AND gate. The output of this is one of the inputs to an OR gate whose other input, 'C' represents the 'r' of our earlier model of the same situation.

We see here that *either* (A *and* B) *or* C will cause an output of 1 at 'Y' in the diagram.

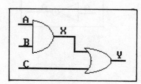

Figure 14.7 *And* combined with *or*

As a truth table this will closely resemble the earlier version but once again the symbolism will have changed:

Table 14.9

A	B	(A.B)	C	(A.B)+C
1	1	1	1	1
1	1	1	0	1
1	0	0	1	1
1	0	0	0	0
0	1	0	1	1
0	1	0	0	0
0	0	0	1	1
0	0	0	0	0

Such models are the very essence of design strategies for complex switching applications of computers and are often 'hard-wired logic' systems.

14.6.10 There is no limit to the complexity of switching circuits but we do, quite often, need to INVERT an input or an output to conform to our logical model.

Suppose, for example, that we have as a statement, "If in A and *not* in B then DO ......". Somehow we have to be able to input 'NOT B', commonly written as $\overline{B}$, and the 'gate' used for this purpose is symbolised as in Figure 14.8, where we see the 'A' input transformed into the 'not A' or $\overline{A}$, also written, you may recall, as A'.

Figure 14.8   Invert gate

14.6.11 When we follow this a little further we may see circuits of the sort of complexity of Figure 14.9 where the combination of two 'inverts', three 'ands' and one 'or' may, at first sight, seem quite bewildering.

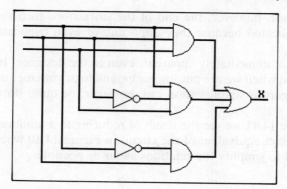

Figure 14.9 A complex of gates

14.7 Analysing Circuits

14.7.1 One way of trying to understand better what is happening is to label all the paths through the diagram in Figure 14.10.

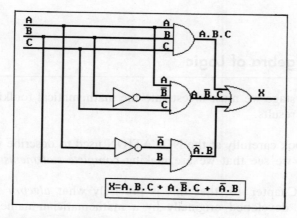

Figure 14.10 The diagram when labelled

14.7.2 If we follow each input, through any 'inverters' if present, to each of the gates in turn, we can see that each output from the corresponding 'and' gate can then also be written.

By definition each of these outputs will be a '1' or a '0' and *they* serve as the inputs to the 'or' gate which, as we saw earlier, is represented by the '+' sign.

14.7.3 Reading the outputs from the 'and' gates, from the top down, and combining them by the 'or' gate, we see the final output to be a combination of all the individual inputs to the final stage.

14.7.4   This is not, however, the end of the story since such switching circuits are often over-complicated because they *duplicate*, or even triplicate or quadruplicate, functions.

Often this is not immediately apparent, even to the designer. It is sometimes even less obvious to *us*, when we are putting compound logic statements into our programs. The results of rigorous simplification can, however, be quite startling.

14.7.5   In Figure 14.11 we see the result of reducing to a minimum the gates needed to provide the exact equivalent of the circuit in Figure 14.10 when the rules of strict logic are applied to simplify the relations as far as possible.

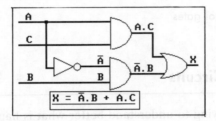

Figure 14.11   Figure 14.10 simplified

14.8   The Algebra of Logic

14.8.1   At this stage we must investigate the mathematical toolkit which allows us to achieve such results.

14.8.2   If we look carefully at the NOTATION used to describe what is happening in Figure 14.10 we see that we are making *complete statements* whilst using *only symbols*.

We saw, in Chapter 10, that this is precisely what *algebra* is and there is a special algebra, developed originally by a mathematician named George Boole, which precisely fits our needs.

14.8.3   This BOOLEAN ALGEBRA uses much the same processes as ordinary algebra, especially in *simplification*. It has its own specialised laws, as well as slight variations of those with which we have already become familiar.

The symbolism is actually a good deal simpler since no coefficients of variables are involved, but we do need to remind ourselves that the symbolism of sets and two-state logic include: A; A'; $\overline{A}$, for example, and two operators, '+', as in A+B, for 'or' and, as in A.B, the '.' stands for 'and'.

14.8.4   Using these symbols our original laws, largely the same as those in Chapter 1, may be re-written as:

1a) $A+\overline{A} = 1$	1b) $A.\overline{A} = 0$
2a) $A+0 = A$	2b) $A.1 = A$
3a) $A.0 = 0$	3b) $A+1 = 1$

14.8.5   These first six may be explored using a simple Venn diagram (as in Figure 14.12) on the understanding that '1' represents the whole universe of discourse and '0', in turn, represents the NULL condition.

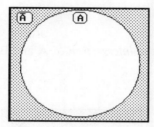

Figure 14.12   One set and the laws

14.8.6   The next pair of laws relate to a two-input, or two-set, compound statement and may, in similar fashion, be explored using a Venn diagram such as that in Figure 14.13.

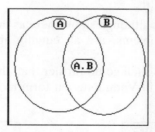

Figure 14.13   Two input diagram

The two laws:
      4a) $A+B = B+A$ and
      4b) $A.B = B.A$

do, however, govern two quite different conditions, the 'OR', the UNION, A+B, in Figure 14.14, together with NOT(A+B).

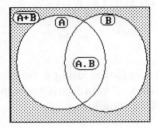

Figure 14.14   A + B and its 'not'

14.8.7   There is, in law 4b) the *intersection* of A and B and Figure 14.15 shows both this and its 'not'.

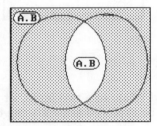

Figure 14.15   The 'and' with its 'not'

14.8.8   The remaining laws, although shown here as three-variable, three-statement, three-set or three-input problems, are all equally applicable to multiple variable models.

In this book, however, we shall go no further than three-component combinations, two of which are illustrated in Venn diagram form in Figures 14.16 and 14.17.

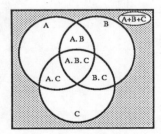

Figure 14.16   A + B + C and its 'not'

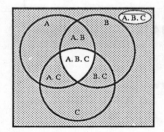

Figure 14.17   A.B.C and its 'not'

14.8.9 The laws, from 5 onwards are:

5a) A+(B+C) = (A+B)+C
5b) A.(B.C) = (A.B).C
6a) A.(B+C) = (A.B)+(A.C)
6b) A+(B.C) = (A+B).(A+C) **

Those marked ** are additional to our original set of laws as are:

7a) A.A = A   7b) A+A = A
8a) A.(A+B) = A   8b) A+(A.B) = A

with one other addition, law 9, $(\overline{\overline{A}}) = A$.

These laws govern our use of the tools by which we state and simplify the Boolean algebra of complex logic systems, much of which can be achieved using the algebraic methods we considered earlier.

14.8.10 Look back at our example in Figures 14.10 and 14.11, above, and let us use the laws, step-by-step to simplify what we had originally:

X = A.B.C + A.$\overline{B}$.C + $\overline{A}$.B which may be simplified as:
X = A.C.(B) + A.C.(B) + A.B(Law 5b)
X = A.C.(B+$\overline{B}$) + $\overline{A}$.B   (Law 6a)
X = A.C.1 + $\overline{A}$.B   (Law 1a)
X = A.C + A.B

which corresponds to what we saw in Figure 14.11.

14.9 De Morgan's Laws

14.9.1 Usually referred to separately are two further laws, known as De Morgan's:

10a) $(\overline{A+B}) = \overline{A}.\overline{B}$ 10b) $(\overline{A.B}) = \overline{A}+\overline{B}$

If, in addition to our 'ordinary' Boolean algebra, De Morgan's laws are used in conjunction with the others in any simplification process, we must follow a strict sequence:

1   Convert every AND to OR and every OR
     to AND.

2   INVERT all individual 'variables', so
     that, for example A becomes $\overline{A}$ and $\overline{B}$
     becomes just B.

3   INVERT the whole final model.

14.9.2 As always the best way to grasp what is happening is to take an example and follow it through the process and then to get plenty of practice with a number of examples.

Let us look at the statement $\overline{A}.\overline{B} = \overline{\overline{A}.B} \equiv \overline{A.B}$

At first sight this seems very odd since the first part on the right-hand side appears to un-balance the equation.

Strict logic is, however, not always immediately obvious to the inexperienced eye and we need to go through the procedure quite deliberately in order to show whether such a statement is true.

Applying De Morgan's laws, we have, after step 1:

$$A+B. \ (\overline{A+B}) \equiv (\overline{A+B})$$

and after step 2 this becomes:

$$A+B. \ (\overline{\overline{A}+\overline{B}}) \equiv (\overline{A}+\overline{B})$$

and, finally, $A+B \ . \ A+B \equiv A+B$

14.9.3 The problem may be solved also by comparing truth tables. The whole statement $\overline{A}.\overline{B} + \overline{A}.B \equiv \overline{A.B}$, may be looked at as a left-hand side, which produces Table 14.10:

Table 14.10

A	B	$\overline{A}$	$\overline{B}$	$\overline{A}.B$	$\overline{A.B}$	$\overline{A} \ . \ \overline{B}$	$\overline{A}.\overline{B}+\overline{A} \ . \ B$
1	1	0	0	0	1	0	0
1	0	0	1	0	0	1	1
0	1	1	0	0	0	1	1
0	0	1	1	1	0	1	1

whereas the right-hand side produces:

Table 14.11

A	B	A.B	$\overline{A}$. $\overline{B}$
1	1	1	0
1	0	0	1
0	1	0	1
0	0	0	1

Since the right-hand column of each table is identical, the *sequence* of statements must generate identical results and the equivalence is demonstrated.

14.10 Karnaugh Maps

14.10.1 There is yet another, much-used, tool for simplifying complex expressions in Boolean algebra or complex switching circuits, especially if they contain numerous 'OR' gates (disjunctions).

Once again, a mathematician is remembered in the name of this device, the Karnaugh map. Rather like Venn diagrams these maps allow us to show all possible combinations of up to six inputs, or variables.

14.10.2 The simplest, for two inputs, is rather like a cross between a matrix and a Venn diagram. It is organised in rows and columns as is a matrix but each 'cell', rather than being a place for storing a variable, is a kind of 'truth domain' representing the combinations of the As and Bs.

In Figure 14.18 we see the 'framework' for such a map which has been filled with the relationships corresponding to each domain in turn. These are ordinarily left blank and if, in a compound statement, that particular relation is present, a tick is put into the corresponding square of the map.

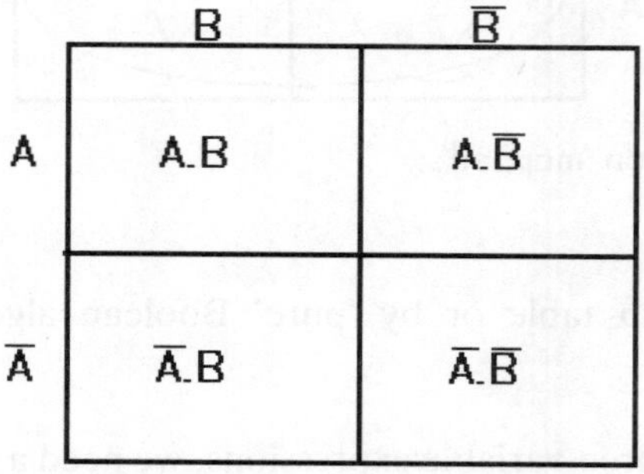

Figure 14.18 The map's domains

The other preliminary step is to learn the domains of the map and two of the four are illustrated in Figures 14.19 and 14.20.

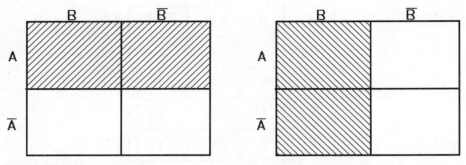

Figure 14.19   The 'A' domain   **Figure 14.20**   The 'B' domain

14.10.3   Let us now look at the process of simplifying a compound statement, such as A.B + $\overline{A}$.B + $\overline{A}$.$\overline{B}$.

We first tick the AB square, next we tick the $\overline{A}$.B square and, finally, the $\overline{A}$ $\overline{A}$.$\overline{B}$ square.

Next we put a loop around each pair of squares, containing 'ticks' (or crosses for that matter) which touch each other along any one side.

As we saw in Figures 14.19 and 14.20 each pair of such squares is one domain of the map; in our example we have 'looped' the $\overline{A}$ and the B domains and our complete expression can be written as $\overline{A}$ + B, in Figure 14.21.

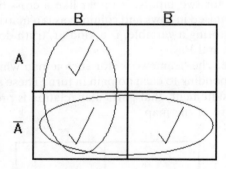

Figure 14.21   The expression 'mapped'

14.10.4   Check, by truth table or by 'pure' Boolean algebra, that this result is correct.

14.10.5   If we move to three-variable expressions, we need a rather more complicated map which is really a cylinder, of all things!

This is a rather more complicated idea but is one which we must remember since such maps are normally drawn as two-dimensional but, since they are really continuous, have a kind of 'carry round the back' trick to remain true to the continuous model. The small gap, in Figure 14.22, is to remind us that, even if the cylinder is 'split' here, in order to flatten it out to two dimensions, any such split is quite artificial.

We have an exactly similar situation when a flat map of our spherical world is spread out on to a two-dimensional piece of paper, this is also a wrap-around model.

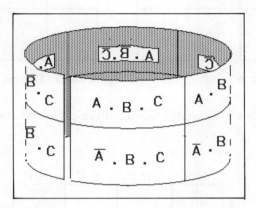

Figure 14.22 The cylinder map

14.10.6 The only other real complication of the three-variable model is that we 'loop together' not only *pairs* of 'cells' but also any blocks of four and any single 'ticked' squares which touch no others in which there are ticks. Have a look at Figure 14.23.

	B.C	B.C̄	B̄.C̄	B̄.C
A	A.B.C	A.B.C̄	A.B̄.C̄	A.B̄.C
Ā	Ā.B.C	Ā.B.C̄	Ā.B̄.C̄	Ā.B̄.C

Figure 14.23 The three-variable map

14.10.7   The domains are different, as we can now see and when we shade the three regions representing A, B and C we see that all of these consist of *four* 'touching' squares as in Figure 14.24 and 14.25.

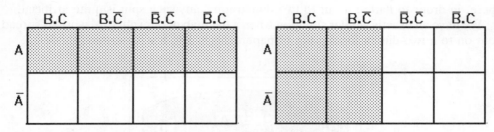

Figure 14.24   Domain A   **Figure 14.25**   Domain B

The C domain follows the wrap-around pattern explained above and appears as in Figure 14.26.

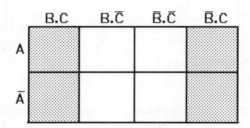

Figure 14.26   The C domain

14.10.8   Can you identify the domains for NOT A, NOT B and NOT C respectively?

14.10.9   We now have the means of simplifying quite complex statements by assigning each compound of *and*s to its appropriate square of the map.
Let us take, as a first example, the complex model:

$$A.B.C + A.B.\overline{C} + \overline{A}.B.\overline{C} + \overline{A}.\overline{B}.C$$

The first compound of *and*s, A.B.C warrants a tick in the top left square of the map; the second, A.B.$\overline{C}$, needs a tick in row 1, column 2 – and so on.

14.10.10   Once all four terms are represented in the model by ticks (as in Figure 14.27) we may draw the loops and start the process of extracting the MINIMUM MODEL, that is to say the representation of the original *complex* of statements by the fewest possible individual elements.
In our case the first loop, row 1 columns 1 and 2, is wholly in the A domain *and*

the B domain and so is A.B. The vertical loop, in column 2, is wholly in the B and the NOT C domains and is, therefore, $B.\bar{C}$ whilst the single circle is in the NOT A and NOT B and C domains and the whole expression is simplified into $A.B + B.\bar{C} + \bar{A}.\bar{B}.C$.

You should, as always, verify this result by one of the other methods we have studied.

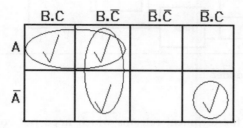

Figure 14.27 Three loops

14.10.11 $A.B.C + A.B.\bar{C} + A.\bar{B}.C + A.B.C + \bar{A}.\bar{B}.C$ looks even more complex but produces only two loops, the one shown as two rough semi-circles being the complete C domain whilst the other is, as in the previous example, A.B, which resolves the whole complex of statements into $A.B + C$. See Figure 14.8.

This is the kind of result which so well demonstrates the power of this final tool in the two-state logical world. We need to become familiar enough with these different techniques to be able to use them in our design and programming and to select the one which best suits our purpose at the time.

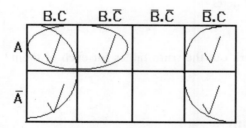

Figure 14.28 The wrap-around

14.11 Half adders

14.11.1 As a matter of some practical interest the gates which add numbers in our computers combine into what have come to be known as 'half adders'.

14.11.2 The results which we need are: 0+0=0; 0+1=1; 1+1=0 carry 1 and, as a block diagram, the circuit is that shown in Figure 14.29.

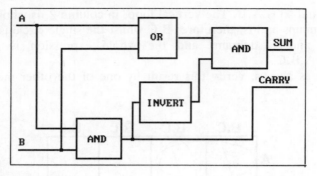

Figure 14.29   The half-adder

14.11.3   In practice, for technical reasons which need not concern us here, AND and OR gates are much less frequently used than NOT AND, NAND, and NOT OR, NOR, gates which invert the outputs of AND and OR gates respectively.

14.11.4   The symbols are shown in Figure 14.30.

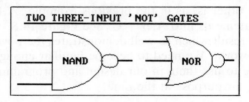

Figure 14.30   The 'not' gates

14.11.5   Can you work out the truth tables for each of them?

14.12   Conclusion

The world of strict logic makes considerable demands upon our thinking but there are few aspects of computing in which a feeling of competence brings greater satisfaction.

Exercises

1) Giving the first statement in each case the designation 'p', the second 'q' and the outcome 'r', write the truth table for:

    i)   Kuala Lumpur is in Malaysia or 3*7 = 21
    ii)   Kuala Lumpur is in Malaysia or 3*7 = 37
    iii) Kuala Lumpur is in Britain or 3*7 = 21
    iv) Kuala Lumpur is in Britain or 3*7 = 37

2) Build up the truth table for $\sim(p \wedge \sim q)$.

3) If 'p' stands for "It is windy" and 'q', "It is raining" write simple sentences to express:

    i) $\sim p$ ii) $p \wedge q$ iii) $p \vee q$ iv) $q \vee \sim p$

    v) $\sim p \wedge \sim q$ vi) $\sim\sim q$

4) If 'p' represents "Hugo is happy" and 'q', "Hugo is rich", (assuming for the moment that both are absolute states), write, in symbolic form:

    i)   Hugo is poor but happy;
    ii) Hugo is neither rich nor happy;
    iii) Hugo is either rich or unhappy;
    iv) Hugo is either poor or is both rich and unhappy.

5) If each of the following pairs of bit sequences were presented to an 'and' gate (as in Figure 14A) write the sequence of bits which would appear at the output:

    i)   110001
         101101

    ii)   10001111
         00111100

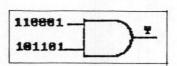

Figure 14A

    iii) 101100111000
         000111001101

6) Show the outputs of a 'not' gate which processes the sequences:

  i)   110001   ii) 10001111   iii) 101100111000

7) Find a Boolean expression and a truth table for the circuit in Figure 14B.

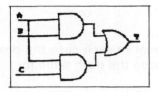

Figure 14B

8) For the logic circuit in Figure 14C:

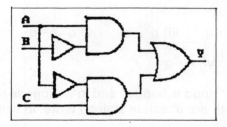

Figure 14C

  a) What inputs will give a 1 at the output?

  b) Write the truth table for the circuit.

9) a) Draw the logical equivalent of the switching circuit in Figure 14D.

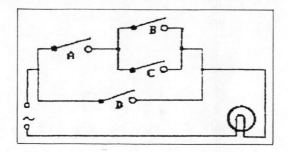

Figure 14D

  b) Give the Boolean expression for the logic circuit.

  c) By Boolean algebra reduce it to its simplest form.

10) Draw the Karnaugh maps and reduce to simplest form the following statements:

  i) A = x.y + x.y'
  ii) A = x.y + x'.y + x'.y'
  iii) A = x.y + x'.y'
  iv) A = x.y.z + x.y.z' + x'.y.z' + x'.y'.z
  v) A = x.y.z + x.y.z' + x.y'.z + x'.y'.z
  vi) A = x.y.z + x.y.z' + x'.y.z' + x'.y'.z + x'.y'.z'

11) For the logic circuit shown in Figure 14E:

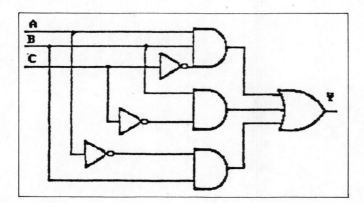

Figure 14E

  a) Write the Boolean expression.

  b) Draw the Karnaugh map and simplify, if possible.

9) a) Draw the logical equivalents of the switching circuit in Figure 14C

Figure 14C

b) Give the Boolean expression for the logic circuit.

c) By Boolean algebra reduce it to its simplest form.

10) Draw the Karnaugh maps and reduce to simplest form the following statements:

  i) $A = \bar{x}y + xy$
  ii) $A = xy + \bar{x}y + \bar{x}\bar{y}$
  iii) $A = xy + \bar{x}y$
  iv) $A = xyz + \bar{x}yz + xy\bar{z} + x\bar{y}z$
  v) $A = x\bar{y}z + xyz + \bar{x}y\bar{z} + x\bar{y}z$
  vi) $A = \bar{x}yz + xyz + \bar{x}\bar{y}z + \bar{x}y\bar{z} + x\bar{x}\bar{y}z$

11) For the logic circuit shown in Figure 14E

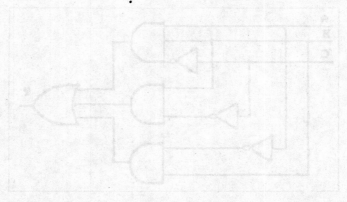

Figure 14E

a) Write the Boolean expression.

b) Draw the Karnaugh map and simplify if possible

Answers to the Exercises

NOTE. Many of the questions set will demand that you make a judgement or give weight to the user's needs, as is quite normal in our working situation. Very occasionally there may be no solution, as in real practice. Where you identify such a situation it is prudent to note your reasons.

It is rarely the case that there will be only one way of performing a calculation and there are often several, even many, solutions to a particular problem. Within these answers what is given will often be a specimen solution but where relevant the essential features or steps will be pointed out. You are asked to check your own answer against the one given and to ensure that, if it does not follow the same path, it arrives at a similar solution or at least contains the minimum elements which are stated.

In the case of models, including pseudo-code versions, names of variables and constants should be stated and the model tested, with appropriate go/no-go data, to ensure that it would serve the purpose.

Chapter 1

1) The laws which do not apply are:
   1a and 1b since $a-b \neq b-a$ and $a/b \neq b/a$;
   2a and 2b; $(5-3)-2 \neq 5-(3-2)$ and $(6/2)/3 \neq 6/(2/3)$;
   3, 4a, 4b and 6 (since $n/0$ is not possible in real terms).

2) a) INTEGER if stated in years or months only but REAL if given as years and fractions of years.
   b) LOGICAL, for example 1 = male, 0 = female or YES/NO respectively; or possibly INTEGER if we use, say, 1 and 2 respectively.
   c) Month – INTEGER – $1 \leq m \leq 12$
      Day – INTEGER – $1 \leq d \leq 31$
      Year – INTEGER – $1900 \leq y \leq 2010$, say. This is a matter of defining a range acceptable to the user and it could be much narrower.
   d) Hour – INTEGER $-0 \leq h \leq 24$ or ≤ 12 if 12-hour clock is used.
      Minute – INTEGER $-0 \leq m \leq 60$
      Second – decided in consultation with the user according to task. Could be INTEGER, $0 \leq s \leq 60$, if for everyday use but if precision timing is involved, in scientific, process-monitoring or athletic records, for example, it would be

REAL and could, in science or computer technology, for instance, involve very small fractions.

e)   REAL, +ve only, $0 \leq$ price $\leq$ top of user's price range. OR may be validated against a pricing file.

f)   INTEGER if wholly converted to cents, pence etc. but REAL if expressed in the normal form $99999.99. For many purposes, especially in the control of error, it is better to use the integer form.

g)   REAL in the range $0 \leq h \leq 168$.

3)   There are many possible answers but any set must pass the tests:

a)   Is it defined? That is to say is there any possible doubt whether an item belongs in or out of the set? If doubt exists the collection is not a set. For general purposes of argument we can accept a less rigorous standard but for strict, two-state logic, there can be no doubt.

b)   Is each element unique? If any element is replicated the collection may be a 'bag' but it is not a set.

4)   Test as in 3), above, or get someone else to apply the same tests. Above all argue the case for and against each example. Many of the things which we meet in our working experience will not be easily identifiable as sets and we should accustom ourselves to testing by the rules.

5)   Many possible answers, e.g. "I must not arrive at work later than 8a.m.", or, "To eat in this restaurant I must have at least enough money to pay the bill."

6)   Many possible answers but we should include:

a)   Validate against the variable type, e.g. a NUMERIC/ALPHA user code.

b)   Verify the range.

c)   Validate against a password file.

d)   Check that the code maps on to the password, character by character.

e)   If not valid then "access denied" message.

Chapter 2

1)   REM let TSV = Total Stock Value; REM open STOCKVALUE file
READ TSV
REM open STOCKFILE
REM open TRANSACTION file
    DOUNTIL EOF
        READ Stocknumber, Quantity, In/Out
        If In/Out = "OUT" then Quantity ← Quantity * −1
        READ from STOCKFILE – key = STOCKNUMBER, Price, Qty. Held
        Qty.Held ← Qty.Held+Quantity
        Value = Price * Quantity

```
        TSV ← TSV + Value
        PRINT Stocknumber, Quantity, Value
        WRITE to STOCKFILE – key = STOCKNUMBER, Qty. Held
    ENDDO
WRITE to STOCKVALUE file, TSV
CLOSE ALL
END
```

2) The model must have inserted into it an outer loop of the type:

```
For P = 1 to 6
NEXT P
```

and must have an array of players' positions subscripted p_1, p_2, and so on.

3) Between the present lines: WRITE S_i and i ← i+1
   we should need to insert a line such as:

   if i>3 then $f_i = f_{i-2}/f_{i-1}$

4) A model on the lines of:

   Let 'd_i' = the distance travelled in one month by a lorry then:
   Cost = $((700+500) + (2*d_i))/d_i$. (Average \$cost/km for 1 month)

   and perform this for each vehicle for each month for a year, say, in order to accumulate the data for an overall 'average' cost per km.

5) a) Employee's contribution: $T_e = 0.01 *$ Gross pay
      Employer's contribution: $T_m = 1.5 * T_e$

      Can be combined as :
      Tax = $(1.5 * (0.01 *$ Gross$)) + (0.01 *$ Gross$)$
          = $(0.015 *$ Gross$) + (0.01 *$ Gross$)$
          = $(0.015 + 0.01) *$ Gross – the distributive law
          = $0.025 *$ Gross
      or T = $(25 *$ Gross$) / 1000$ if we wish to work in integers as far as possible.

   b) Within the payroll loop, after calculating each individual's Gross Pay and *before* any other deductions are made we should insert something like:

      Tax$_i$ = $(25 * G)/1000$
      Total Tax Payable ← Total Tax Payable + Tax$_i$

   c) Gross Pay must be in the range $0 \le GP \le \$5000$, say, or whatever maximum pay may be earned within the user's organisation.

6) Starting record number = S = RND(10), E = S
   REM open PAYROLL file
       DOUNTIL EOF
           GET record number E,
           READ GROSS PAY
           If GROSS PAY < 0 OR GROSS PAY > 1000 then DO ERROR
           E ← E+10
       ENDDO
   ERROR routine
   END ERROR
   Close all
   END

7) This is quite demanding and should be taken in stages, taking, say, the A to E table first and then adding the E to A table.

   a) A pseudo-code model should contain something like this:

   REM Let S = average speed; d_i = the distances between individual stations and T = the time past midnight, 00.00hrs., in minutes, ARRi = the arrival time at a station and DEP_i = the departure time.

   S = 40, T = 420, d_1 = 10, d_2 = 8, d_3 = 14, d_4 = 16
       DOUNTIL T = 1140
       i = 2, DEP1 = T
           DOUNTIL i = 5
               $ARR_i = DEP_{i-1} + ((d_i/S) * 60)$
               IF i < 5 then $DEP_i = ARR_i + 2$
   REM PRINTFILE AtoE
               IF i = 2 then WRITE DEP_1
               WRITE ARR_i
               IF i ≤ 4 then WRITE DEP^i
               i ← i + 1
           ENDDO
       i = 4, DEP_5 = T
           DOUNTIL i = 1
               $ARR_i = DEP_{i+1} + ((d_i/S) * 60)$
               IF i > 1 then $DEP_i = ARR_i + 2$
   REM PRINTFILE EtoA
               IF i = 4 then WRITE DEP_5
               WRITE ARR_i
               IF i > 1 then WRITE DEP_i
               i ← i−1
           ENDDO
       T ← T + 20
       ENDDO
   Close all
   END

This will need to be followed, before printing the tables, by a routine which will convert the times, in minutes past midnight, to the normal hours and minutes form. For each time recorded and INT(T/60) will convert the hours and a T(mod60) routine will give the minutes component. Conversion of the hours to 12 or 24 hour format would then be at the user's discretion.

b) The dry-run of the program, with the results in tabular form, will show us that the first train in either direction will arrive at its destination only at 08.20. Allowing the turn-around time it cannot start its return journey until the 08.40 service by which time FOUR other trains will have started from the terminus. There will thus be FIVE trains, in all, in each direction before return workings are possible. To provide for the one reserve train a total of ELEVEN will be needed to run the services.

Chapter 3

1) a) The set must be defined, ie we *must* be able to say of any item that it is either in or out of the set, there can be no 'perhaps'.
   b) Each element must be unique.

2) There will be an infinite number of possible answers to this question! The only way to check your own is to ask a friend to apply the two tests in Question 1), above.

3) b) e) f) and h) clearly fail the tests in Question 1). Item d) is a very common problem to business database users because it embodies no less than three attributes and all must be the subject of testing questions. "Is 'X' a customer; does he/she live in Hong Kong; is that a sole residence?" are the questions which expose the problem. The customer who has two addresses or has a business in one place but resides in another often needs very careful definition if the file is not to become a bag. It would be far safer to treat it as doubtful than to say categorically that it is a set as it stands.

4)

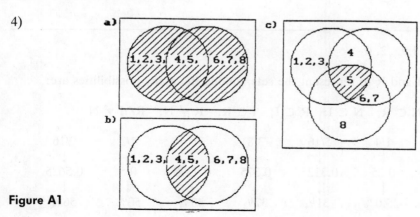

Figure A1

5)

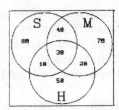

Figure A2

6) Using the symbol, → , to signify 'maps onto':

  a) x → x³
  b) x → x + 32
  c) The names of the months in the Gregorian calendar → their respective lengths
      in days.
  d) x → 2x + 2
  e) x → 20 – x

7) a) The set of males over 1.5m tall
  b) The set containing all the elements of sets A, B and C.
  c) The set containing all the elements which are not, at the same time, in all
      three sets A, B AND C.
  d) The set of elements which are in EITHER A OR B but NOT both.
  e) The set of elements which are NOT in either A OR B *or* in A and B
      combined.
  f) The set of values from −10 to +10, inclusive, but which are not defined as
      REALS or INTEGERS and may, therefore, be either of these.

Chapter 4

1) 4.2, −5, 14.9, $\sqrt{2}$, 0

2) Using N, I and R to represent the sets, among the many possibilities are:

$$N \subset R; \quad N \subset I; \quad R \subset I; \quad I \subset R; \quad R \neq N; \quad R \cap I \neq N$$

3)

3/16	1/4	5/16	3/8	7/16	1/2	9/16
↓	↓	↓	↓	↓	↓	↓
0.1875	0.25	0.3125	0.375	0.4375	0.5	0.5625
↓	↓	↓	↓	↓	↓	↓
18⅜%	25.0%	31¼	37½	43¾	50%	56¼

5/8	11/16	3/4	13/16	7/8	15/16	1
↓	↓	↓	↓	↓	↓	↓
0.625	0.6875	0.75	0.8125	0.875	0.9375	1.0
↓	↓	↓	↓	↓	↓	↓
$62\frac{1}{2}$	$68\frac{3}{4}$	75	$81\frac{1}{4}$	$87\frac{1}{2}$	$93\frac{3}{4}$	100%

The differences in decimal values are 0.0625 and in percentages $6\frac{1}{4}\%$

4) $\{\sqrt{2}, \sqrt{3}, \sqrt{5}, \sqrt{6}\}$

5) $\{1, 2, 3, 4, 5, 6, 7, 8, 9\}$

Chapter 5

1)   3476

2) $30{,}000{,}000 - 400{,}000 = 29{,}600{,}000$ but,
     $300*10^5 - 4*10^5 = 296*10^5$

3)   a)   167
     b)   3E80
     c)   7DA0
     d)   8BD

4)   a)   1199
     b)   413
     c)   17967
     d)   65535

5)   a)   0.349
     b) $285*10^2 = 28500$
     c)   34.9A Hex
     d)   2AF3 Hex

6)   a) 8;   b) 6;   c) 4;   d) $(10+7+8)_{mod11} = 3$

Chapter 6

1)   There are many possible sources of mistakes at this stage among which are:
     document omitted or misread; transposition errors; characters omitted;
     mis-keying; repeated keys due to keyboard 'dwell' or 'bounce. Check your
     answer by exchanging with a fellow-student perhaps.

2) Again there are many possible applications such as controlling the completeness of a batch by hashing the document numbers, eg invoice numbers, by hashing dates, batching totals of amounts of money, total number of items withdrawn from stores – and so on. Check with a colleague or fellow-student.

3) Parity checking is testing the make-up of the binary digits in a coded eight-bit byte by making the leftmost digit a 1, if the total number of 1's in the remaining seven bits is *odd* and even parity is being used or, in odd parity, if the total number of 1's is even. For example:

10110111 would show that the byte is correct if *even* parity is being used but it would show that there had been some corruption if the parity were odd.

4) In each case we multiply each digit in turn by its column weight to give:

Example	Hash Total	MOD7 Value	MOD7 Complement	Final Number
a)	43	1	6	4396
b)	107	2	5	279465
c)	18	4	3	300003
d)	112	0	7	1058537
e)	208	5	2	174937265

5) a) NOT;   b) NOT;   c) NOT;   d) NOT;   e) VALID

6) a) ±0.5km   b) ±0.05cm   c) ±0.005gm   d) ±0.5cm

7) a) ±1.4%   b) ±0.37%   c) ±0.038%   d) ±0.0038%

8) a) The maximum error will be generated if all the top line elements are multiplied together, then all the bottom line elements and finally the division is performed.
   b) i) Error will be diminished if the sequence is: 3.5/7; 13.5/4.5; 1.5/40.5 giving 1/2; 3/1 and 1/27 = 1/18
      ii) In the computer we could consider converting all to integers before performing b)ii).

9) The *reported* answer would be, on the face of it, 25.375

   a) 3*7 = 21. Relative error against the reported answer −15.8%
   b) i) 25.2 if values are cut to 1D: E_r = −0.7%
      ii) 25.4 if only the answer is cut: E_r = +0.1%

c) 3.5*7.3 = 25.55 and E_r = +0.7%

d) 25.4 and E_r = +0.1%

e) Maximum = 25.75525; minimum = 24.99525

(error bound = ± 0.005 on each original value)

Error bound on the answer is approx ±0.38 and $E_r \approx$ ±1.6%

10) If 'r' is not an integer we should, if possible, take its value to 5D and do the same for Pi. If practicable we should then raise 'r' to the third power, multiply the result by Pi, multiply by 4 at the next to last step and finally divide by 3. Only by taking such precautions to minimise error could we genuinely claim to be able to state the result to 3D.

Chapter 7

1) a) 437   b) 001   c) 899   d) 77

2) a) (499 + 633) − 1000 leads us to:

$$\begin{array}{r} 499+ \\ 633 \\ \hline \text{cut } 1|132 \end{array}$$

and, by 'cutting' the overflow:

b) (499 + 363) − 1000, similarly gives: 862 − 1000 = 138

c) (2495 + 9208) − 10000 gives: cut 1|1703

d) (1207 + 9350.5) − 10000   cut 1|0557.5 = 557.5

3) a) 0100 1001 0101
    b) 0010 0010 0010
    c) 0110 0011 0010
    d) 0100 0001 0000

4) In BCD addition if the result of adding two digits is a legal BCD code, ie, 1001 or less then the result stands. If the sum is an *illegal* code, ie 1010 or over the *add* 0110 and carry 0001.

a)
```
0100   1001   0011  +
0010   0111   0101
0110   10000* 1000   *greater than 1001 so add 0110 and carry 1
0001c  0110
0111   0110   1000  = 768₁₀
```

b) By the same process the answer will be: 0010   0001   1000

c) 0001   0100   1001

d) 0001   0100   0100   0011. NOTE that there is a carry into the new fourth
                                        column

5) The point to note in this question is that the 4-bit digits are as they would appear in a partially calculated answer and have not had the 'add 0110 and carry' routine performed. This must be done where necessary to arrive ate the appropriate answers:

a) 7319   b) 1014   c) 5055   d) 7622

6) The key to this question is to remember the 'add 1 to the rightmost place' to convert the one's complement to two's complement form.

a) 0 0 1 1 1 0 1 0
    1 1 0 0 0 1 0 1 – one's complement
                 1+
    1 1 0 0 0 1 1 0 – two's complement

b) 11010100   c) 10110111   d) 11011011

Note: a useful check is to convert the two's complement form back into denary and when added to the original number the total should always be 256. Can you see why this should be so?

7) a) i) 63 as the exponent and 0.11111111 as the mantissa.
       ii) −63 as the exponent and 0.00000001 as the mantissa.
   b) i) 63 as the exponent and 0.111111111111111111111111 as the mantissa.
       ii) −63 as the exponent and 0.000000000000000000000001 as the mantissa.

8) We are able to instruct the machine to calculate to higher precision and in this way to minimise internal error propagation. This can be achieved only at the cost of considerably slower processing.

9) Using lines to show the separation, which does not actually occur in the machine, we have: 1100010001111000

Chapter 8

1) a)

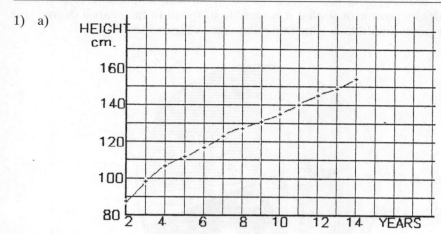

Figure A3 A child's height

1) b)

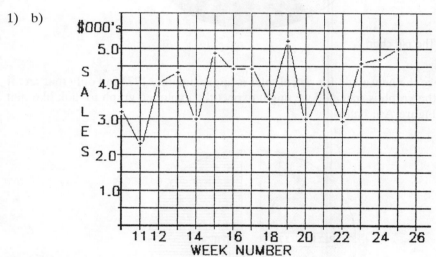

Figure A4 The sales graph

2) a) The graph in Figure A3 shows a more rapid growth rate from 2 to 4 years and relatively steady growth thereafter.

   The graph in Figure A4 shows that sales fluctuated considerably from month-to-month.

   b) One might reasonably interpolate in Figure A3 from 6 to 14 years but there is little practical value in doing so. The extreme fluctuations of Figure A4 make it quite useless for interpolation.

3) a)   The angles would be, approximately, Petrol 169.4° Repairs 95.3°; Car Hire
         5.4° and Car Sales 89.9° but it would be virtually impossible to plot such
         angles by protractor so that it is sensible to round to the nearest degree and
         the graph should look something like that in Figure A5.

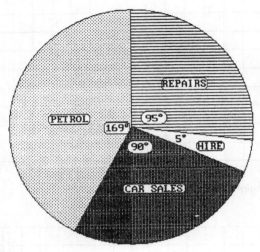

Figure A5   The pie chart

3) b)   Since the new angle for petrol sales is approximately 177° and only that sector
         of the graph is specified we might expect the exploded graph to look like that
         in Figure A6, below.

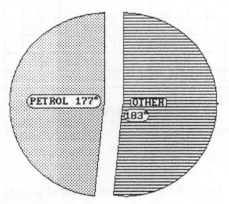

Figure A6   The exploded pie

3) c)   For a direct comparison we first convert to percentages of the two, three-
         month period, totals and find that in the first petrol sales accounted for
         47.05% of the total and 49.03% of the total for the second period. The
         increase is, therefore, approximately 2%.

4) A histogram has a continuous scale as a baseline, normally on the horizontal axis, as a consequence of which the area of each segment is significant. It is truly a two-dimensional graph. The block or column chart has a discontinuous scale on one axis and only the other dimension is significant.

5) a) Table of moving averages:

Year	3	4	5	6	7	8	9	10	11	12	13	14
Avge. Ht.	92.5	99.8	105.9	111.4	117.2	122.1	126.6	130.8	135.4	140.2	144.6	149.3

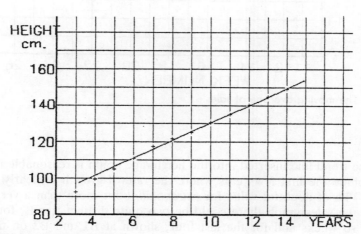

Figure A7   Moving average height/age

5) a) continued) The table for the moving average of sales is :

Week	11	12	13	14	15	16	17	18	19	20	21	22	23	24	25
Avge. Sales	2.75	3.75	3.84	3.42	4.11	4.25	4.33	3.96	4.58	3.79	3.89	3.39	3.99	4.34	4.67

and the corresponding graph:

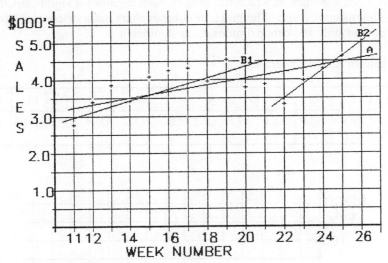

Figure A8   The moving average of sales

5) b) The trend line inserted into the height/age graph is reasonable if it does not include the first few years where the rate of growth is clearly different. It will be of doubtful value for *extrapolating* for more than a very few years. In the sales graph there could be two quite distinct trends, for the first 20 or so weeks and for the last four, shown at B1 and B2 on the graph. It would perhaps be unwise to take such a short-term view since fluctuations are known to be considerable and the overall trend line, A on the graph, may well be more reliable. The sales staff might prefer to take line B2 as the trend! Mathematically the evidence is not strong for placing great reliance on any of them.

6) The table of calculated values might look like:

If x =	−4	−3	−2	−1	0	1	2	3	4
+4	4	4	4	4	4	4	4	4	4
$2x^2=$	32	18	8	2	0	2	8	18	32
$2x=$	−8	−6	−4	−2	0	2	4	6	8
then y=	28	16	8	4	4	8	16	28	44

and the corresponding graph will appear as in Figure A9. Note that the graph has been drawn to a very compressed 'y' scale for reasons of space. In practice it is always preferable to make the scales as large as the size of paper will permit and in cases such as this, where the 'y' values are very much larger than the 'x' values, the longer axis should be vertical so as to enhance the accuracy of plotting the graph.

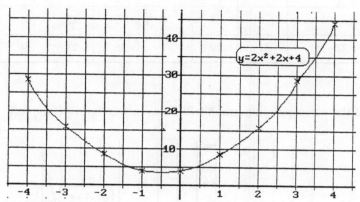

Figure A9   The square-law curve

The effect of the changes in scaling may be seen by comparing Figure A9 with the computer-scaled version in Figure A10.

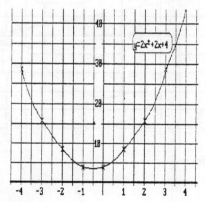

Figure A10   The 'Y' axis scaled

We should always select the scales most carefully and use the largest possible area of graph paper, especially if we may need to interpolate or read off intermediate values from our graph. The other points to remember are to draw a smooth curve, to identify the scales and to give the graph a title.

7) Once again it pays to tabulate the calculations carefully:

a)

If x =	−2	−1	0	1	2	3	4
x³ =	−8	−1	0	1	8	27	64
+4	4	4	4	4	4	4	4
y=x³+4 =	−4	3	4	5	12	31	68
y=27−2x =	31	29	27	25	23	21	19

and the corresponding graph will be as in the figure to the left.

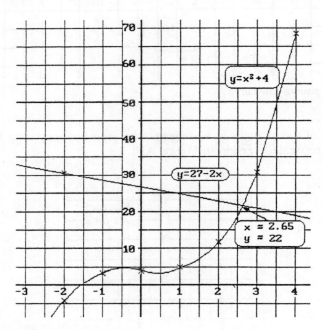

Figure A11   The two functions

b) The point of intersection of the curves on the graph is at x ≈ 2.6 and y ≈ 22. If you have drawn your graph to a larger scale and on a larger sheet of graph paper you *may* be able to reach closer approximations. What would you consider the error bound to be in the drawing of your own graph?

c)

the region defined by the constraints given will appear as in Figure A12.

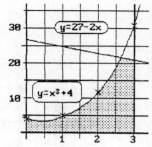

Figure A12   The feasibility region

8) a) If we use 'L' to stand for the number of large boxes and 'S' for the number of small each of the functions may be modelled as:

SALES, $L \le 280$; $S \le 400$
PRODUCTION, $480 \ge 1.5L + 0.75S$ since it takes 1.5 minutes to make a large box and 0.75 minute to make a small.
STORAGE, $4 \ge 0.01L + 0.008S$

b) We can then say, for example that if $L = 0$ production may be, at most, $480/0.75 = 640$ small boxes and that if $S = 0$ production may be at most 320 large boxes. In this way we can calculate the three inequalities as:

SALES: $0 \le L \le 280$; $0 \le S \le 400$
PRODUCTION: if $L = 0$ then $S \le 640$; if $S = 0$ then $L \le 320$
STORAGE: if $L = 0$ then $S \le 500$; if $S = 0$ then $L \le 400$

c) The graph of the feasibility region will be as shown in Figure A13. Within the shaded region, including the lines which make up its boundaries, any variation of large and small boxes is theoretically feasible. The best *practical solution* will, in practice, be decided by the user after taking other demands, such as profitability, into account.

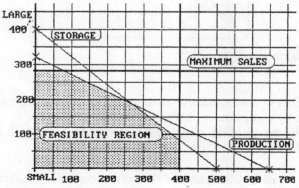

Figure A13   The production feasibility region

Chapter 9

1) The plot should look like that in Figure A14.

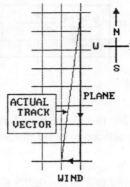

Figure A14   The flight plot

2) a) The plot of the course sailed will look like that in Figure A15.

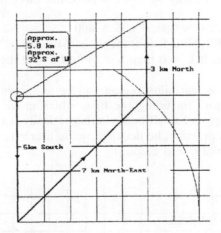

Figure A15   The course sailed

The points to note specially are that the North-East 'leg' must be measured or set out by compasses as here.

Similarly the distance remaining and the direction to be sailed must be measured.

The alternative description of the course back to the starting point, as an ordered pair or single vector is very much simpler.

2) b) The ordered pair is $(-5, -3)$ which represents 5km West and 3km South.

3) The shape $\begin{bmatrix} 1 & 5 & 1 & 2 & 0 \\ -2 & 1 & 1 & 5 & 2 \end{bmatrix}$

is plotted in Figure A16. Note that the points are both labelled, A, B, etc., and indicated by a circle.

This is done for clarity only and has no special significance except that it allows us to follow what is happening when we transform the shape.

In a similar way the scales are numbered on this plot but the numbering will be omitted in later answers for the sake of the clarity of the picture.

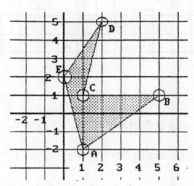

Figure A16 The shape plotted

4) a) $\begin{bmatrix} 3 & 7 & 3 & 4 & 2 \\ 1 & 4 & 4 & 8 & 5 \end{bmatrix}$ leads us to Figure A17

in which we see that the addition of the vector leads to the translation of the shape in both the 'x' and the 'y' axes.

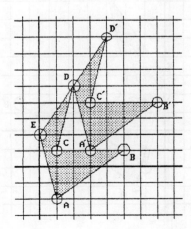

Figure A17 'S' translated

b) $\begin{bmatrix} -1 & -5 & -1 & -2 & 0 \\ -2 & 1 & 1 & 5 & 2 \end{bmatrix}$ is the next transform which leads to a reflection of the shape as in Figure A18.

4) b) i)

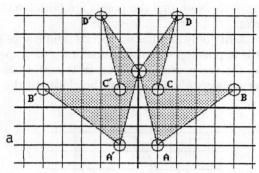

Figure A18   A reflection

ii) the transform is a reflection in the 'y' axis.

4) c) All the values will be multiplied by the scalar, 1.5 to give:

$$\begin{bmatrix} 1.5 & 7.5 & 1.5 & 3 & 0 \\ -3 & 1.5 & 1.5 & 7.5 & 3 \end{bmatrix}$$

and this in turn will produce the shape in Figure A19.

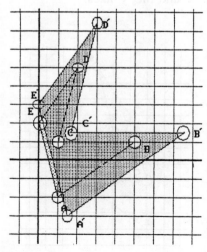

Figure A19   The enlarged shape

5) There will always be many possible ways of programming any matrix operation but one of the most direct ways is by using nested loops with the outer loop, the 'i' counter, matching the number of rows and the inner, 'j' loop, corresponding to the number of columns.

If we designate the 'subject' matrix as 'S', the transform vector or matrix as 'T' and the resultant matrix as 'R' we can set up some permanent models.

For a *scalar* multiplication:

For i = 1 to r; rem 'r' = number of rows
   For j = 1 to c; rem 'c' = number of columns
      $r_{i,j} = s_{i,j}$ * Scalar; rem '$r_{i,j}$' = the individual cell of the *resultant* matrix,
rem $s_{i,j}$ = the individual cell of the subject matrix.
      Next j
Next i

NOTE: the middle line will contain the '+' operator for a scalar addition.

For a vector addition, adding $\begin{pmatrix} v_1 \\ v_2 \end{pmatrix}$ to the subject matrix, S the 'core' line of the model would be:

$$r_{i,j} = s_{i,j} + v_i$$

For a multiplication of, say, a 2x5 matrix by a 2x2 matrix, as in several of our examples, the core line would be:

$$r_{i,j} = (s_{1,j} * t_{i,1}) + (s_{2,j} * t_{i,2})$$

Any transform involving two or more stages will involve putting together two or more of these operations into the same model.

Chapter 10

1) If we use 'M' for the matrix representing the machines, 'S' for the sub-assemblies, 'C' for the component costs, 'P' for the prices of assemblies and 'T' for the total cost per machine, the operations will be as follows:

a) With machines A, B and C as the columns and sub-assemblies 'd' and 'e' as the rows:

$$M = \begin{bmatrix} 3 & 2 & 3 \\ 3 & 4 & 5 \end{bmatrix} \quad S = \begin{bmatrix} 3 & 2 \\ 4 & 3 \\ 3 & 5 \end{bmatrix} \text{ with sub-assemblies 'd'}$$

with sub-assemblies 'd' and 'e' as the columns and 101a, 101b and 101c as the rows whilst the cost matrix, with 101a, 101b and 101c as the columns will be:
$C = (3\ 2\ 4)$

b) The cost of each machine, in the matrix T, will then be the prices of sub-assemblies, P = C*S, times the makeup of machines matrix, 'M' leading to:

$$(3\ 2\ 4)* \begin{bmatrix} 3 & 2 & 3 \\ 3 & 4 & 5 \end{bmatrix} = P = (29\ 32)$$ as the price of assemblies 'd' and 'e' respectively. P*M is then:

$$(29\ 32) * \begin{bmatrix} 3 & 2 & 3 \\ 3 & 4 & 5 \end{bmatrix} = \$(183\ 186\ 247) = T$$ for machines 'A', 'B' and 'C' respectively.

c) The total cost of one days output will be:

i) $\begin{pmatrix} 80 \\ 50 \\ 40 \end{pmatrix}$ for the numbers produced multiplied by 'T'

ii) $(183\ 186\ 247) * \begin{pmatrix} 80 \\ 50 \\ 40 \end{pmatrix} = \$33820$

iii) the new cost vector 'C$_2$' = (3.15 2.15 4.16) so the new assemblies price will be 'P$_2$' = (30.53 33.55)

P$_2$ * M will give T$_2$ = (192.24 195.26 262.25)

and $T_2 * \begin{pmatrix} 80 \\ 50 \\ 40 \end{pmatrix} = \$35632.20$, an increase of \$1812.20.

2) In most high-level languages arrays must be declared as dedicated to specific variable types. They may hold any legal variable types but not a mixture. In practice we very commonly need to store, in complete records, mixtures of variable types and linked lists allow us to do this since each list may be associated with a distinct variable type but the association *between lists* is not variable-dependant.

3) Lists are linked so that records may be held in any required order, or may be ordered as needed on any individual field without the integrity of any record being jeopardised. For example a company's customer file may, at different demands, be sorted alphabetically by customer name, say, or numerically by customer number, without the records becoming jumbled.

4) i) A pointer is essentially the continuity indicator of any list, that is to say that it may point to another of the fields which make up the single record, to where the next item in a list is to be found, to where the previous item is held or to the next in sequence of a sorted list.

ii) Your answer should be a sketch which resembles that shown in Figure 10.2 of Chapter 10 in this book.

5) In a 'LIFO' stack the 'pushing' or 'popping' of a single item does not disturb the rest of the stack whereas in a 'FIFO' stack the whole stack has to be displaced or the complete set of pointers changed when an item is either pushed or popped.

6) There are many possible answers here and the only sure way to test your own solution is to do a pencil-and-paper dry run or to write a short program segment and test-run it with suitable data.

7) Here too there are many possible illustrations. The essential features, however, are that the 'sample' number should always be compared first with the mid-value of the range available, the 'test' value, in our case 75. If the sample number is greater than the test number that the range should be *halved* with the lower half 'discarded'. If, on the other hand, the test value is larger than the sample the upper half should be discarded. If they are equal the search must show a successful end.

In our example the first test value will be 75, the second 112 (or 113), the third 94, the fourth 84, the fifth 79, the sixth 77. Only six passes will be needed for our search.

8) a) Your answer should closely resemble the tree in Figure A20.

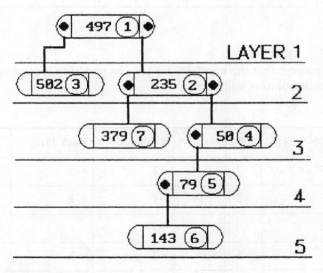

Figure A20 The data tree

b) The table, with its pointers will be:

NODE	CONTENTS	L.Pointer	R.Pointer
1	497	3	2
2	235	7	4

NODE	CONTENTS	L.Pointer	R.Pointer
1	497	3	2
2	235	7	4
3	502	−1	−1
4	50	5	−1
5	79	6	−1
6	143	−1	−1
7	379	−1	−1

c) If we assume that the numbers are seen as sorted in *ascending order* the trace and back pointers will be:

NODE	CONTENTS	L.Pointer	R.Pointer	Back Ptr.	Trace Ptr.
1	497	3	2	−1	3
2	235	7	4	1	7
3	502	−1	−1	1	4
4	50	5	−1	2	5
5	79	6	−1	4	6
6	143	−1	−1	5	2
7	379	−1	−1	2	1

d) 128 would have to fit in as the right pointer from Node 6 and the table would be:

NODE	CONTENTS	L.Pointer	R.Pointer	Back Ptr.	Trace Ptr.
1	497	3	2	−1	3
2	235	7	4	1	7
3	502	−1	−1	1	4
4	50	5	−1	2	5
5	79	6	−1	4	8
6	143	−1	8	5	2
7	379	−1	−1	2	1
8	128	−1	−1	6	6

9) a) Assuming that the matrices represent three places, 'A', 'B' and 'C' and that the matrices are in sequence with columns 'A', 'B', 'C' and rows 'A', 'B' and 'C' respectively, the networks would be as in Figure A21. NOTE: row 'A' connected to column 'A', for example, must be shown in some fashion similar to the loops in Figure A21.

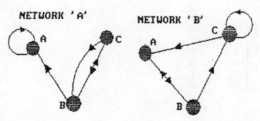

Figure A21 Networks A & B

b) The matrices would sum to:

$$\begin{bmatrix} 1 & 1 & 0 \\ 2 & 0 & 2 \\ 1 & 2 & 1 \end{bmatrix}$$

and the network will be as in Figure A22:

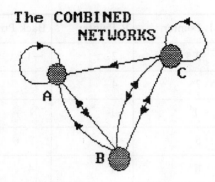

Figure A22   Networks A & B combined

c) The links which are one-way only are the sole path from A to B and one path from A to C. They are most easily identified if arrows are always put on to the relevant tracks.

10) a) If, as in question 9, we label the rows from 'A' to 'D', downwards, and the columns from 'A' to 'D', from the left, only row 'D' and column 'D' are connected to all of the other stations. The server must, then, be position 'D'.

b) If each station must be linked, by two-way links, to each of the others the relevant matrix will be:

i)
$$\begin{bmatrix} 0 & 1 & 1 & 1 \\ 1 & 0 & 1 & 1 \\ 1 & 1 & 0 & 1 \\ 1 & 1 & 1 & 0 \end{bmatrix}$$

ii) We can find the difference between this and what exists at present by subtracting the present matrix from the proposed matrix:

$$\begin{bmatrix} 0 & 1 & 1 & 1 \\ 1 & 0 & 1 & 1 \\ 1 & 1 & 0 & 1 \\ 1 & 1 & 1 & 0 \end{bmatrix} - \begin{bmatrix} 0 & 1 & 0 & 1 \\ 0 & 0 & 1 & 1 \\ 0 & 0 & 0 & 1 \\ 1 & 1 & 1 & 0 \end{bmatrix} = \begin{bmatrix} 0 & 0 & 1 & 0 \\ 1 & 0 & 0 & 0 \\ 1 & 1 & 0 & 0 \\ 0 & 0 & 0 & 0 \end{bmatrix}$$

and this tells us what new paths will be needed.

Chapter 11

1) Estimate range   Next estimate   Result**

1.25	$<\sqrt{2}<$	1.5	$\frac{1}{2}(1.25+1.5)$	$= 1.375$	$<\sqrt{2}$
1.375	$<\sqrt{2}<$	1.5	$\frac{1}{2}(1.375+1.5)$	$= 1.4375$	$<\sqrt{2}$
1.375	$<\sqrt{2}<$	1.4375		$= 1.40625$	$<\sqrt{2}$
1.40625	$<\sqrt{2}<$	1.4375		$= 1.421875$	$<\sqrt{2}$

and so on to:

$1.4140625 <\sqrt{2}<$ 1.414306641 and $\sqrt{2} \approx 1.414$ to 3D

since the difference between the two roots which make the 'estimate range' is too small to affect the rounding to 3D.

Note that this takes a further *eleven* iterations and that the 'RESULT'** is established by *squaring* the next estimate to find if it is greater or less than the root we seek.

2) Your model may take many forms in detail but it must contain:
i) a loop; ii) a test to refuse negative numbers; iii) a test to see if the new estimate is greater or less than the value we seek; iv) a procedure to take the previous *smaller* estimate if the new value is too large or the *larger* previous estimate if the new is too small; v) a 'break-out' routine which will test for an acceptable error bound between each pair of consecutive estimates, for example ' IF $|e_i - e_{i-1}| < 0.0001$ THEN END.

3) a) The model must accept REAL numbers as input and must reject NEGATIVE NUMBERS.

   If e_i is any given estimate, $e_1 = \frac{1}{2}x$ and $e_2 = x/e_1$ then:
   $e_{i+1} = \frac{1}{2}((x/e_i) + e_i)$
   and $y = e_i$ when $|e_i - e_{i-1}| <$ acceptable error.

   b) i)   for answers to 3D the acceptable error must be ≤ 0.0001;
      ii) for answers to 4 sig. figs. we would have to insert a manual 'stop' to the iteration since the range of acceptable 'x' values is such as to make the number of places in the root quite unpredictable.

4) At the seventeenth pass we should arrive at:

$21.679 < \sqrt{470} < 21.682$ and $\sqrt{470} \approx 21.68$ to 2D

or after twenty passes we could write the answer as 21.679 to 3D.

5) You should have found, fairly quickly, that this is a very slow convergence indeed and that it is impractical to perform it by hand. After writing a short program to

perform it, which took several thousand passes, I found the value of π, to 3d, to be 3.142, a difference of 0.001 or $E_r = 0.03\%$

6) $R = (P+T)/H; H = (P+T)/R; T = (R*H)-R$

7) a) $3x^3 + 3x^2 - 2 = y$

b) If

x =	1	2	3	4
$3x^3 =$	3	24	81	192
$+3x^2 =$	3	12	27	48
$-2 =$	-2	-2	-2	-2
Then y =	4	34	106	238

8) If

$$y = 2x^2 + 27$$

$$y - 27 = 2x^2$$

$$\frac{1}{2}(y-27) = x^2$$

$$x = \sqrt{\frac{y-27}{2}}$$

9) a) $6t + 4s = 54$

$6t + 15s = 153;$ $11s = 99;$ $s = 9,$ $t = 3;$

b) The matrix of the left-hand side of the equations becomes:

$\begin{bmatrix} 3 & 2 \\ 2 & 5 \end{bmatrix}$ and its INVERSE is: $\dfrac{1}{15-4}\begin{bmatrix} 5 & -2 \\ -2 & 3 \end{bmatrix}$

$\dfrac{1}{15-4}\begin{bmatrix} 5 & -2 \\ -2 & 3 \end{bmatrix} * \begin{pmatrix} 27 \\ 51 \end{pmatrix} = \begin{pmatrix} 135 - 102 \\ -54 + 153 \end{pmatrix} = \begin{pmatrix} 33 \\ 99 \end{pmatrix}$ and $\dfrac{1}{11}\begin{pmatrix} 33 \\ 99 \end{pmatrix} = \begin{pmatrix} 3 \\ 9 \end{pmatrix}$

giving the result, t = 3, s = 9.

This appears to be a more laborious method but, as is explained in the chapter, the matrix method is easy to program and can work for *any* values whereas the 'traditional' method depends upon inspection and is quite difficult to program.

10) $\dfrac{x^5 * x^{-1}}{x^{-2} * x^4} = \dfrac{x^{5-1}}{x^{-2+4}} = \dfrac{x^4}{x^2} = x^{4-2} = x^2$

Chapter 12

1) The mean is 25.

2) a)

a_i	x_i	f_i
$26 \leq 30$	28	5
$31 \leq 40$	35.5	109
$41 \leq 50$	45.5	168
$51 \leq 60$	55.5	188
$61 \leq 70$	65.5	105
$71 \leq 80$	75.5	14
$81 \leq 90$	85.5	1

b) i)

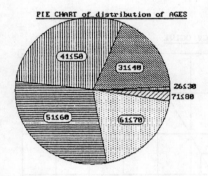

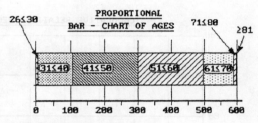

Figure A24   The bar-chart

Figure A23   The pie chart

ii)

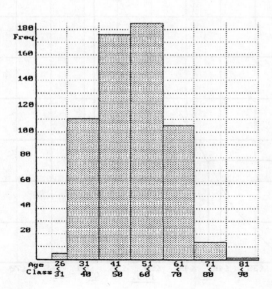

Figure A25   The histogram

2) b)   CLASS   Cu. f   Cu. f %
    i)   ≤30   5   0.83
            ≤35   48   8.0
            ≤40   114   19.0
            ≤45   205   34.2
            ≤50   292   48.7
            ≤55   397   66.2
            ≤60   480   80.0
            ≤65   560   93.3
            ≤70   585   97.5
            ≤75
            ≤80   599   99.8
            ≤95   600   100

ii) The Ogive should be drawn with as smooth a curve as is possible, as in Figure A26 below.

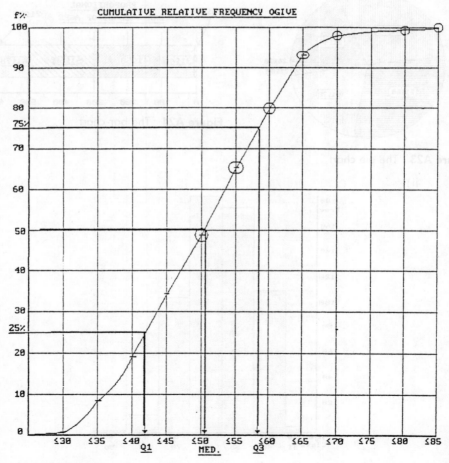

Figure A26   The relative cu. f. ogive

iii) We can also see from the table that the MODE is 53 and we can read the MEDIAN and the SIQR from the graph, below, as approximately 51 and ½(58-42) or approximately 8 respectively.

c) By the standard procedures the descriptive statistics, taken from the original table, are:

MEAN ≈ 50.6 VARIANCE ≈ 108.8 S.D. ≈ 10.4

NOTE that all of these should be stated as 'years' if there is any possibility of their being confused with any other data.

3) We should calculate and plot the moving 'average' and look to the line of best fit to indicate a trend.

NO. The evidence is over too short a period and there could be many possible explanations for such variations. It could be that any one month could be better than average, because of gift-buying on a large scale, say, or similarly that a month could be relatively short, for example February in the Gregorian calendar.

Such a picture could very well be a reason for seeking further evidence but is not, of itself, sufficient to lead to any firm conclusions.

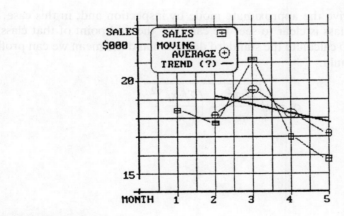

Figure A27 The trend in sales

4)

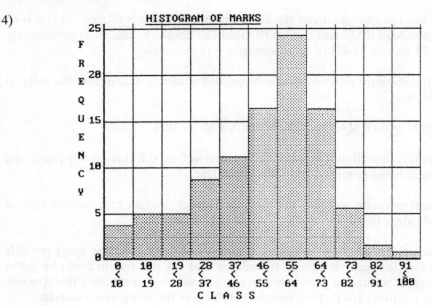

Figure A28

a) The histogram should look like the one on the left and should have a title, scales on both axes and no gaps between the 'columns'.

b) We may derive the approximate mode by inspection and, in this case, since the largest class is clear to see, we can take the mid-point of that class, 59.

   In order to calculate the standard deviation and the mean we can profitably use the formula:

$$\sigma = \sqrt{\frac{\Sigma f_i x_i^2}{n} - \left(\frac{\Sigma f_i x_i}{n}\right)^2}$$

but in order to do so we can benefit from making a data table:

CLASS	X_1	x_1^2	f_1	f_1x_1	$f_1x_1^2$
0 < 10	5	25	3	15	75
10 < 19	14	196	5	70	980
19 < 28	23	529	5	115	2645
28 < 37	32	1024	8	256	8192
37 < 46	41	1681	12	492	20172
46 < 55	50	2500	17	850	42500
55 < 64	59	3481	24	1416	83544
64 < 73	68	4624	17	1156	78608
73 < 82	77	5929	6	462	35574
82 < 91	86	7396	2	172	14792
91 < 100	95	9025	1	95	9025
Σ			100	5099	296107
MEAN				50.99	

and substituting the relevant values into the formula we have:

$$\sigma = \sqrt{\frac{296107}{100} - \left(\frac{5099}{100}\right)^2} = \sqrt{2961.07 - 2599.98} = \sqrt{361.12} = 19.00$$

and Mean = 51.00, Mode = 59

c) In this case the range concerned would be from $32 \leq x \leq 70$ and, by apportioning between the classes at the ends of this range we can arrive at a figure of approximately 69 who have scored in the ±1 s.d. range. The rule holds good here.

d) By the same means we can find that *outside 2.s.d. from the mean*, includes those scoring less than 13 or more than 89 and in our distribution we can find about 7 such candidates, or 7% of the whole. This is nearly double the 4% and this distribution does not conform to that rule.

5) a) See graph - Figure A29.

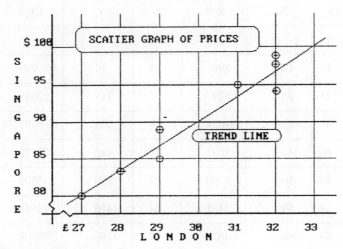

Figure A29 Singapore/London compared

b) The scatter is tightly concentrated around the line marked on the graph and this would indicate that the prices in the two centres *are* closely related to one another.

6) a) These data are rather awkward to organise into an odd number of classes with a whole number as the mid-point of each. Seven classes of an interval of 2½ is about the best we can manage and this distribution generates a mean of 67.35 a mode of approximately 67 and a median of 67.8. The S.D. is approximately 3.1.

b) There is little to be gained from anything more than a histogram.

Chapter 13

1) a) 4845
   b) 1/5
   c) 3/190
   d) 32/625

2) a) 5/22
   b) Yes. There are 5 programmers in all of which you must be one so there are only four others out of a total of 21, if you are the 'onlooker'. The change in probability is small but it *does* exist.

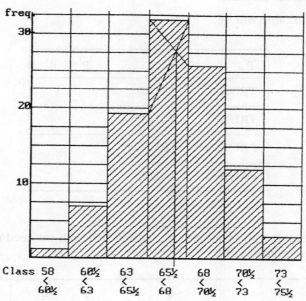

Figure A30   The histogram — drawn 'mode'

3) a) There are 114, out of 600, who are ≤ 40 so this probability is 114/600 or 19/100, (19%).

b) 1/100

c) 188 are between 50 and 60 and roughly 1/10 of those between 40 and 50 might be expected to fit the requirement. Thus approximately 205 out of 600 or 41/140 is the probability we seek.

4) Standard scores are found by the model (Score - Mean)/S.D. so:

a) Maths $(80-48)/16 = 2 = z_m$
Programming $(65-55)/12 = 0.83 = z_p$

b) i) 2.3% and approximately 22.1% respectively (by interpolation)

ii) Maths 2.3% of 60 = 1 person (since we cannot have fractions!)
Programming 13 people.

iii) 2.3/100 * 22.1/100 $\approx$ 50/10000 or 1/200

Chapter 14

1) a) Here the 'or' is not defined as 'xor' and we must conclude that the statements will return a '1' if either or both propositions are true.

The truth table will be:

'p'	'q'	'p' or 'q'
TRUE	TRUE	1
TRUE	FALSE	1
FALSE	TRUE	1
FALSE	FALSE	0

2) This involves three propositions, 'p', 'q' and '~q' and we need more columns to construct the original table:

'P'	'q'	'~q'	'p' ∧ '~q')	~('p' ∧ '~q')
TRUE	TRUE	FALSE	FALSE	1
TRUE	FALSE	TRUE	TRUE	0
FALSE	TRUE	FALSE	FALSE	1
FALSE	FALSE	TRUE	FALSE	1

3) i) It is not windy ii) It is windy and raining iii) It is windy or it is raining iv) It is windy or it is not raining v) It is not windy and it is not raining vi) It is not true that it is not raining - it is raining.

4) i) ~p∧q;   ii) ~p∧~q;   iii) p∧~q;   iv) ~p∨(p∧~q)

5) An 'AND' gate only returns a '1' at the output when both inputs are '1'.

So:

i) 100001   ii) 00001100   iii) 000100001000

6) i) 001110   ii) 01110000   iii) 010011000111

7) A.B + A.C + A.B.C. is the Boolean expression since it is an 'OR' gate. The truth table will be:

A	B	C	Y
0	0	0	0
0	0	1	0
0	1	0	0
0	1	1	0
1	0	0	0
1	0	1	1
1	1	0	1
1	1	1	1

8) A.B̄ + Ā.C with the truth table:

A	B	C	Y
0	0	0	0
0	0	1	1
0	1	0	0
0	1	1	0
1	0	0	1
1	0	1	0
1	1	0	0
1	1	1	0

9) a) The logic diagram, using gates, will be:

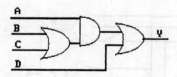

Figure A31

b) A.(B+C) + D

c) This expression cannot be simplified.

10)

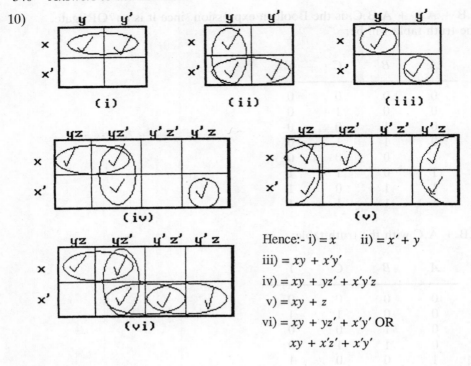

(i)   (ii)   (iii)

(iv)   (v)

(vi)

Hence:- i) $= x$ ii) $= x' + y$

iii) $= xy + x'y'$

iv) $= xy + yz' + x'y'z$

v) $= xy + z$

vi) $= xy + yz' + x'y'$ OR

$xy + x'z' + x'y'$

Figure A32

11 a)   By labelling the outputs of each gate in turn we get Figure A33:

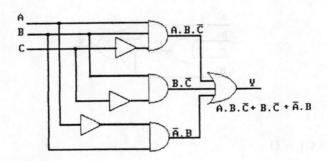

Figure A33   The gates labelled

from which we can see that the output from 'y' is any one, two or three of
the outputs from the respective 'and' gates.

b) When we put this into a Karnaugh map, as in Figure A34.

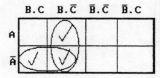

Figure A34   The map

we can see that all the components are contained within the rectangle consisting of:

A.B.C̄ + Ā.B.C̄ + Ā.B.C.

Index